Lecture Notes in Computer Science 9511

Commenced Publication in 1973
Founding and Former Series Editors:
Gerhard Goos, Juris Hartmanis, and Jan van Leeuwen

More information about this series at http://www.springer.com/series/7407

Ioannis Karydis · Spyros Sioutas
Peter Triantafillou · Dimitrios Tsoumakos (Eds.)

Algorithmic Aspects of Cloud Computing

First International Workshop, ALGOCLOUD 2015
Patras, Greece, September 14–15, 2015
Revised Selected Papers

 Springer

Editors
Ioannis Karydis
Department of Informatics
Ionian University
Kerkyra
Greece

Spyros Sioutas
Ionian University
Corfu
Greece

Peter Triantafillou
University of Glasgow
Glasgow
UK

Dimitrios Tsoumakos
Ionian University
Kerkyra
Greece

ISSN 0302-9743   ISSN 1611-3349   (electronic)
Lecture Notes in Computer Science
ISBN 978-3-319-29918-1   ISBN 978-3-319-29919-8   (eBook)
DOI 10.1007/978-3-319-29919-8

Library of Congress Control Number: 2016931342

LNCS Sublibrary: SL1 – Theoretical Computer Science and General Issues

Printed on acid-free paper

This Springer imprint is published by SpringerNature
The registered company is Springer International Publishing AG Switzerland

Preface

The International Workshop on Algorithmic Aspects of Cloud Computing (ALGO-CLOUD) is an annual event aiming to tackle the diverse new topics in the emerging area of algorithmic aspects of computing and data management in the cloud. The increasing adoption of cloud computing introduces a variety of parallel and distributed algorithmic models and architectures. To leverage elastic cloud resources, scalability has to be a fundamental architectural design trait of new cloud databases. This challenge is manifested in new data models (NoSQL), replication, caching and partitioning schemes, relaxed consistency and transaction guarantees, as well as new protocols, APIs, indexing and storage services.

The aim of the workshop is to bring together researchers and practitioners in cloud computing algorithms, service design, and data architectures to exchange ideas and contribute to the development of this exciting and emerging new field.

ALGOCLOUD welcomes submissions of theoretical, experimental, methodological, as well as application papers. Demonstration papers and high-quality survey papers are also welcome. As such, contributions are expected to span a wide range of algorithms for modeling, constructing, and evaluating operations and services in a variety of systems, including (but not limited to) virtualized infrastructures, cloud platforms, datacenters, mobile ad hoc networks, peer-to-peer and grid systems, HPC architectures, etc.

Topics of interest addressed by this workshop include, but are not limited to:

- Algorithmic aspects of elasticity and scalability for distributed, large-scale data stores (e.g., NoSQL and columnar databases)
- Search and retrieval algorithms for cloud infrastructures
- Monitoring and analysis of elasticity for virtualized environments
- NoSQL, schemaless data modeling, integration
- Caching and load-balancing
- Storage structures and indexing for cloud databases
- New algorithmic aspects of parallel and distributed computing for cloud applications
- Scalable machine learning, analytics, and data science
- High availability, reliability, failover
- Transactional models and algorithms for cloud databases
- Query languages and processing, programming models
- Consistency, replication and partitioning CAP, data structures and algorithms for eventually consistent stores

ALGOCLOUD 2015 took place during September 14–15, 2015, at the Conference and Cultural Center of the University of Patras, Greece. It collocated and was part of ALGO 2015, the major annual congress that combines the premier algorithmic conference European Symposium on Algorithms (ESA) and a number of other specialized conferences and workshops, all related to algorithms and their applications, making

ALGO the major European event for researchers, students, and practitioners in algorithms.

ALGOCLOUD 2015 was organized by the University of Patras and its Department of Computer Engineering and Informatics, and it was supported by Springer as well as by the European Social Fund (ESF) complemented with Greek national funds through the Operational Program Education and Lifelong Learning of the National Strategic Reference Framework (NSRF) – Research Funding Program THALES: Investing in Knowledge Society through the European Social Fund.

The Program Committee (PC) of ALGOCLOUD 2015 was delighted by the positive response to the call for papers. The diverse nature of papers submitted demonstrated the vitality of the algorithmic aspects of cloud computing. All submissions underwent the standard peer-review process and were reviewed by at least three PC members, sometimes assisted by external reviewers. The PC decided to accept 13 original research papers that were presented at the workshop.

The program of ALGOCLOUD 2015 was complemented by two highly interesting tutorials. The first one, entitled "Performance and Scalability of Indexed Subgraph Query Processing Methods," was delivered by Prof. Peter Triantafillou (School of Computing Science, University of Glasgow, UK). The second tutorial, entitled "Distributed Privacy Preserving Record-Linkage," was delivered by Prof. Vassilios Verykios (School of Science and Technology, Hellenic Open University, Greece). We wish to express our sincere gratitude to both distinguished scientists for the excellent tutorials they provided.

We hope that these proceedings will help researchers to understand and be aware of state-of-the-art algorithmic aspects of cloud computing, and that they will stimulate further research in the domain of algorithmic approaches in cloud computing in general.

September 2015

Ioannis Karydis
Spyros Sioutas
Peter Triantafillou
Dimitrios Tsoumakos

Organization

Program Co-chairs

Spyros Sioutas   Ionian University, Greece
Peter Triantafillou   University of Glasgow, UK
Dimitrios Tsoumakos   Ionian University, Greece

Program Committee

Christos Anagnostopoulos   University of Glasgow, UK
Alexis Delis   University of Athens, Greece
Marios Dikaiakos   University of Cyprus, Cyprus
Schahram Dustdar   Technical University of Vienna, Austria
Anastasios Gounaris   AUTH, Greece
Seif Haridi   Royal Institute of Technology, Sweden
Ioannis Karydis   Ionian University, Greece
Yannis Konstantinou   NTUA, Greece
Christos Makris   University of Patras, Greece
Haralambos Mouratidis   University of Brighton, UK
Nikos Ntarmos   University of Glasgow, UK
George Pallis   University of Cyprus, Cyprus
Mema Roussopoulos   University of Athens, Greece
Hong-Linh Truong   Technical University of Vienna, Austria

External Reviewers

Marios Kendea
Andreas Kosmatopoulos
Nikolaos Nodarakis
Athanasios Naskos

Abstracts

Performance and Scalability of Indexed Subgraph Query Processing Methods

Peter Triantafillou

School of Computing Science, University of Glasgow, UK

Abstract. Graph data management systems have become very popular, as graphs are the natural data model for many applications. One of the main problems addressed by these systems is subgraph query processing; i.e., given a query graph, return all graphs that contain the query. The naive method for processing such queries is to perform a subgraph isomorphism test against each graph in the dataset. This obviously does not scale, as subgraph isomorphism is NP-complete. Thus, many indexing methods have been proposed to reduce the number of candidate graphs that have to underpass the subgraph isomorphism test. In this tutorial, we identify a set of key factors–parameters that influence the performance of related methods: namely, the number of nodes per graph, the graph density, the number of distinct labels, the number of graphs in the dataset, and the query graph size. We then discuss comprehensive and systematic experiments that analyze the sensitivity of the various methods on the values of the key parameters. Our aims are twofold: first to draw conclusions about the relative performance of the algorithms, and, second, to stress-test all algorithms, deriving insights as to their scalability and highlighting how both performance and scalability depend on these factors. We present six well-established indexing methods, namely, Grapes, CT-Index, GraphGrepSX, gIndex, Tree+, and gCode, as representative approaches of the overall design space, including the most recent and best-performing methods. We report on their index construction time and index size as well as on query processing performance in terms of time and false-positive ratio. We discuss performance on both real and synthetic datasets. Specifically, four real datasets of different characteristics are used: AIDS, PDBS, PCM, and PPI. In addition, we report on a large number of synthetic graph datasets, empowering us to systematically study the performance and scalability of the algorithms as they depend on the aforementioned key parameters.

A Tutorial on Blocking Methods
for Privacy-Preserving Record Linkage

Dimitrios Karapiperis[1,✉], Vassilios S. Verykios[1],
Eleftheria Katsiri[2], and Alex Delis[3]

[1] School of Science and Technology, Hellenic Open University, Patras, Greece
{dkarapiperis, verykios}@eap.gr
[2] Department of Electrical and Computer Engineering,
Democritus University of Thrace, Xanthi, Greece
eli@imis.athena-innovation.gr
[3] Department of Informatics and Telecommunications,
University of Athens, Athens, Greece
ad@di.uoa.gr

Abstract. In this paper, we first present five state-of-the-art private blocking methods which rely mainly on random strings, clustering, and public reference sets. We emphasize on the drawbacks of these methods, and then, we present our L-fold redundant blocking scheme, that relies on the Locality-Sensitive Hashing technique for identifying similar records. These records have undergone an anonymization transformation using a Bloom filter-based encoding technique. Finally, we perform an experimental evaluation of all these methods and present the results.

Contents

Tutorial

A Tutorial on Blocking Methods for Privacy-Preserving Record Linkage

Dimitrios Karapiperis[1]([✉]), Vassilios S. Verykios[1], Eleftheria Katsiri[2], and Alex Delis[3]

[1] School of Science and Technology, Hellenic Open University, Patras, Greece
{dkarapiperis,verykios}@eap.gr
[2] Department of Electrical and Computer Engineering,
Democritus University of Thrace, Xanthi, Greece
eli@imis.athena-innovation.gr
[3] Department of Informatics and Telecommunications,
University of Athens, Athens, Greece
ad@di.uoa.gr

Abstract. In this paper, we first present five state-of-the-art private blocking methods which rely mainly on random strings, clustering, and public reference sets. We emphasize on the drawbacks of these methods, and then, we present our L-fold redundant blocking scheme, that relies on the Locality-Sensitive Hashing technique for identifying similar records. These records have undergone an anonymization transformation using a Bloom filter-based encoding technique. Finally, we perform an experimental evaluation of all these methods and present the results.

Keywords: Bloom filter · Locality-sensitive hashing · Blocking

1 Introduction

A series of economic collapses of bank and insurance companies recently triggered a financial crisis of unprecedented severity. In order for these institutions to get back on their feet, they had to engage in merger talks inevitably. One of the tricky points for such mergers is to be able to estimate the extent to which the customer bases of the constituent institutions are in common, so that the benefits of the merger can be proactively assessed [33]. The process of comparing the customer bases and finding out records that refer to the same real world entity, is known as the Record Linkage, the Entity Resolution or the Data Matching problem [3]. Record Linkage consists of two steps. In the first step potentially matched pairs are searched while in the second step these pairs are matched. The searching step addresses the problem of bringing together for comparison tentative matched pairs of records, while disregarding the unpromising ones. The searching step should be able to identify a minimal superset of the matched pairs so that no computational resources are wasted in comparison operations during the following step. The second step, known as the matching step, entails the

© Springer International Publishing Switzerland 2016
I. Karydis et al. (Eds.): ALGOCLOUD 2015, LNCS 9511, pp. 3–15, 2016.
DOI: 10.1007/978-3-319-29919-8_1

comparison of record pairs which have been brought together for comparison in the previous step. The matching step is implemented either in an exact or in an approximate manner. An exact matching of two records can be regarded as a binary decision problem with two possible outcomes denoting the agreement or disagreement of these records. Approximate matching comprises the calculation of a continuous value similarity metric that usually assumes values in the range of $[0, 1]$.

When data to be matched is deemed to be sensitive or private, such as health data or data kept by national security agencies, Privacy-Preserving Record Linkage (PPRL) techniques should be employed [11]. PPRL investigates how to make linkage computations secure by respecting the privacy of the data, and imposes certain constraints on the two steps of Record Linkage just described, on the top of the necessary anonymization of the input records. First of all, the anonymization of the records must be implemented in such a way that (a) no sensitive information in a record is disclosed to parties other than the owner, (b) the anonymization process is time and cost efficient (c) it preserves the distance of the values in the record fields, i.e., if record a is closer to record b than it is to record c, then the same relationships should hold for their anonymized counterparts, and (d) the final deliberation about the linking status of a pair of records, that relies on the comparison of their anonymized form, should be a close approximation of the distance between their original record counterparts. The PPRL process is summarized in Fig. 1.

Fig. 1. The PPRL process.

The secure searching solutions, which have been developed to solve the PPRL problem, rely mostly on traditional blocking, where all records that have the same value in a specific field(s) are placed together for comparison. However, the proposed solutions exhibit a considerable overhead in terms of performance, when applied to voluminous data and especially to high-dimensional data. In [13] by Inan et al. as an example, blocking relies on the categorization of records into generalized hierarchies based on the semantics of values of selected fields, which may lead to load imbalance problems, if most values semantically belong to certain categories. Karakasidis et al. in [17] present a blocking technique which relies on a sliding window that creates blocks of records. Its performance is considerably degraded, when the size of that window is increased in order to produce more accurate results. Authors in [7, 19, 21] use redundant probabilistic methods, where each record is blocked to several independent blocking groups, in order to amplify the probability of bringing together similar records for comparison. Authors though utilize an arbitrary number of blocking groups, which has as a result either unnecessary and expensive comparisons or missed similar record pairs.

In this paper, we first present five state-of-the-art private blocking methods which rely mainly on random strings, clustering, and public reference sets. We emphasize on their drawbacks, and then present our new flexible L-fold redundant blocking scheme which is structured around an efficient technique for searching potentially matching record pairs. More specifically, our scheme relies on the idea of blocking one record to multiple groups in order to amplify the probability of inserting similar records into the same block. We use the so-called Locality-Sensitive Hashing technique [9], where we utilize only the necessary number of blocking groups. By doing so, we achieve accurate results without imposing any additional computational overhead. This LSH-based searching method, as shown experimentally in Sect. 5, can reduce the number of record pairs that are brought together for comparison up to 98 % of the total comparison space. Experimental results demonstrate the effectiveness and the superiority of our method by comparing it to five state-of-the-art private blocking methods.

The structure of the paper is organized as follows: Related work is given in Sect. 2. In Sect. 3, we illustrate some basic building components used by the private blocking methods presented in Sect. 4. These methods are evaluated in Sect. 5, while conclusions are discussed in Sect. 6.

2 Related Work

Several solutions have been provided in the literature in the filed of efficient searching for similar records [2,5,7,13–15,17,19,21,26]. However, the proposed solutions exhibit poor performance when applied to large data sets. In [5] for example, a cheap distance metric is used for creating clusters of records and then a more expensive, accurate distance metric is used to evaluate the record pairs that are tagged for further evaluation. Nevertheless, the number of record pairs that should be compared can still be excessively large. Authors in [2] use TF/IDF [25] in order to generate weight vectors from each record, which are used as keys during the blocking mechanism. Various levels of privacy protection are presented at the expense of efficiency. The tree-based indexing methods used in [14,15,26], in order to reduce the number of candidate record pairs, as reported and proved in [10,32] and [1], deteriorate rapidly to quadratic complexity, by scanning the whole index structure repeatedly, when these structures are used for representing records even with moderate dimensionality ($\geq$10). A detailed survey of blocking techniques for Record Linkage can be found in [4]. An overview of privacy-preserving blocking techniques is provided in [31].

Our scheme utilizes a trusted third-party and we make the assumption that this party does not collude with the other participants. Two-party techniques, like the ones in [29,30], may reduce the risk of privacy breach but they are complex and they add high communication cost.

3   Building Components

In this section, we present two basic building components of the blocking methods which will be presented later.

3.1   Secure Multi-party Computations

A reliable Secure Multi-Party (SMC) technique of performing a joint computation among several parties is the partially homomorphic Paillier cryptosystem [24]. A joint computation could be the addition of some values, where these values should remain secret due to privacy concerns. Successive encryption of the same value generates different cipher texts with high probability.

A trusted authority is required in order to issue a public/private key pair, needed for the encryption and decryption operations respectively. Given two values (messages), x_1 and x_2, encryption is performed by using the public key and the produced cipher texts are denoted by $\widetilde{x}_1$ and $\widetilde{x}_2$ respectively. Given the cipher texts, we can perform either homomorphic addition $(\widetilde{x}_1 \oplus \widetilde{x}_2)$ or multiplication with a constant c $(c \odot \widetilde{x}_1)$. The cipher texts can be decrypted by the trusted authority by using its private key. SMC protocols are effective, reliable but add high computational overhead.

3.2   Differential Privacy

Generally, Differential Privacy (DP) [8] aims to maximize the accuracy of queries posed to statistical databases, while minimizing the chances of disclosing the identities of the real-world entities represented therein. DP is achieved (a) by defining the sensitivity of each query, and (b) by adding noise. The sensitivity is the maximum difference obtained by applying a query on two databases which differ on at most one row. These databases are termed as sibling or neighboring databases.

Let us assume, two sibling medical databases which contain records that represent patients. Let us also assume, one boolean attribute which denotes if a patient has a certain disease or not (e.g., diabetes). The output of the count query "How many patients have diabetes?" applied on both databases will differ by at most one, which is the sensitivity S of that query.

A randomized mechanism $\mathcal{M} : \mathcal{D} \to \mathcal{O}$ applied on the sibling databases $\mathcal{D}_1$ and $\mathcal{D}_2$, which are essentially two sets of rows of database $\mathcal{D}$, is differentially private if for all sets $O \subseteq \mathcal{O}$, it holds that

$$\Pr[\mathcal{M}(D_1) \in O] \le e^{\epsilon} \Pr[\mathcal{M}(D_2) \in O], \tag{1}$$

where the probability is taken over the coins of $\mathcal{M}$. In order to achieve Eq. 1, mechanism $\mathcal{M}$ adds to the true value of the query, noise drawn from a zero-mean Laplace distribution with scale S/ϵ, where ϵ is the privacy parameter. By increasing the value of ϵ, we achieve stronger privacy guarantees but we experience accuracy loss in the responses.

4   Private Blocking Methods

In this section, we present six state-of-the-art private blocking methods.

4.1   Method HG

This hierarchy-based blocking method (HG) [13] relies on the categorization of records into generalized hierarchies based on the semantics of values of selected attributes. The authors use k-anonymity [28] to generate these hierarchies which comprise blocks of possible matching record pairs. Generalized attribute values are sent to a third party, who classifies the formulated record pairs as matches, non-matches, or possible matches. Then, an SMC approach is used to calculate similarities of the possible matches. The records of the sample database shown in Table 1 are categorized into hierarchies illustrated in Table 2.

Table 1. Sample database

1	John Smith	Doctor
2	Andy Petterson	Doctor
3	John Smyth	Teacher
4	Susan Devon	Teacher

Table 2. Generalized 2-anonymous database

1	Doctor
2	Doctor
3	Teacher
4	Teacher

A modification of HG, proposed in [14], relies on the notion of DP for providing strong privacy guarantees. In this method, the data custodians independently partition their records into d-dimensional regions (hyper-rectangles) using tree-based indexing structures (i.e., BSP-Tree, KD-Tree, or R-Tree), and then exchange the differentially-private perturbed size and extents of these regions. During the blocking phase, only records which have been partitioned into regions that are compliant to a certain decision rule are considered. For example, record a which belongs to region *[Female/age 40–45]* cannot formulate a pair with record b which belongs to *[Male/age 20–25]*. The generalized hierarchies may lead to load imbalance problems, if most values semantically belong to certain generalized hierarchies.

The main drawbacks are (a) the computationally expensive SMC approach, and (b) the difficulty of applying it on strings that do not semantically belong to a certain generalized hierarchy.

4.2   Method EUC

The Euclidean distance-based blocking method (EUC), was proposed by Scannapiecco et al. [26]. This method represents strings in a private manner by embedding them in a Euclidean space. EUC uses P reference sets, common to both data custodians who participate in the linkage process. In these reference sets, each element comprises a random sequence of characters of length approximately equal to the average length of strings in the data sets. Embedding a string s results in a vector of size P, where each component of this vector stores the minimum edit distance of s from all the elements in a reference set. Figure 2 illustrates the process of building the vectors of two similar string. The authors, in order to find the similar vectors, use a multidimensional tree-based index, which has the performance drawback described in Sect. 2. For this reason, we utilize the Euclidean LSH-based blocking scheme [6,20] specifically developed for finding similar points in Euclidean spaces.

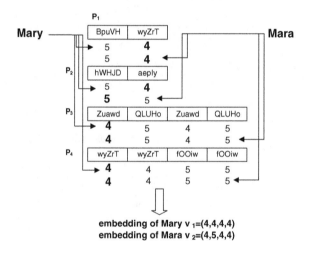

embedding of Mary v_1=(4,4,4,4)
embedding of Mara v_2=(4,5,4,4)

Fig. 2. Building the vectors of strings 'Mary' and 'Mara'.

4.3   Method PHN

The phonetic encoding-based blocking method (PHN), by Karakasidis and Verykios [16], used phonetic encoding functions to generalize strings. For example, using the Double Metaphone encoding method, *'SMITH'* and *'SMYTH'* are both encoded as *'SM0'*. The phonetic algorithms are computationally fast, which makes them appealing to settings with a large number of records. First, the data custodians convert their strings into phonetic codes and additionally inject fake codes into the encoded data sets, which are then sent to a trusted third party. The third-party builds blocks using common codes from both data sets. Next, the matching codes are returned to the data custodians and the corresponding records are exchanged.

This method falls short of representing similar strings effectively due to the inadequacy of the phonetic codes to represent these similar values with the same code.

4.4   Method AHC

Kuzu et al. proposed an agglomerative hierarchical clustering-based blocking method (AHC) [22], which is based on public reference sets and DP. Global clusters are generated by the trusted third-party for a set of public reference values of a chosen field. Each data custodian assigns her records into these clusters according to their similarity with respect to the elements of each cluster. Differential privacy is used in order to perturb the cardinality of each cluster by adding noise randomly drawn from a zero-mean Laplace distribution. The authors add two types of noise, namely (a) positive noise which is incorporated by adding fake records to the blocks, while (b) negative noise requires suppressing original records. This addition/suppression of records, which is illustrated in Fig. 3, entails accuracy loss in the final result set.

Next for the matching step, for the field values found in common clusters, an SMC approach is followed by representing each field value as a binary vector, where each component represents a distinct bigram. A component of that vector is set to 1 if the bigram that represents appears in the corresponding field value. However, these vectors suffer from excessively high dimensionality and also exhibit high degree of sparsity.

The main drawback of AHC is the strong dependence on the public reference sets, which should be a subset of the values used in the data sets (as will be shown in Sect. 5). Additionally, the SMC approach followed is computationally expensive.

4.5   Method TPB

In this two-party blocking method (TPB), introduced in [30], each data custodian creates a separate reference set for a sorting attribute value (e.g., 'Last-Name'), merges her records with it, and sorts the results. The data custodians independently generate clusters, which contain a reference value, and a list of attribute values sorted before that reference value. These clusters are merged such that each cluster contains at least k attribute values in order to achieve k-anonymity [28].

Then, those reference values which correspond to each cluster are first exchanged between the data custodians and then are merged and sorted. We note that the number of the reference values, denoted by n_e, exchanged for each cluster can be 1 or more. This parameter plays an important role on the privacy guarantees of TPB. The sorted nearest neighborhood method [12] is applied on

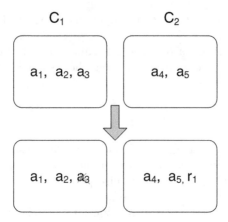

Fig. 3. Suppressing record a_3 in cluster C_1 and adding fake record r_1 in C_2.

the this sorted list of reference values using a sliding window of size w. Eventually, record pairs are formulated by using the reference values, which represent certain clusters, that fall into the same window.

TPB exhibits the same deficiency with AHC, since there is a strong dependence on the choice of the public reference sets. As we will see in the experiments, the reference values should be a subset of the values used in the data sets in order to achieve accurate results. Also, the authors do not mention anything about the protocol/method which will be used for computing the distance between the records of each pair.

4.6   Method HLSH

Method HLSH [20] is based on Hamming Locality-Sensitive Hashing technique [9] and on a Bloom filter-based encoding method [27]. Schnell et al. have shown that Bloom filters are able to preserve the distance between strings. A bitmap array of size S, initialized with zeros, is created by hashing all consecutive bigrams of a string (sequences of pairs of adjacent characters), by using independent composite cryptographic hash functions which may include either *MD5* and *SHA1*, or more advanced keyed hash message authentication code (*HMAC*) functions like *HMAC-MD5* and *HMAC-SHA1*, which utilize a secret key. We concatenate Bloom filters, which represent single field values of a record, in order to compose record-level Bloom filters. Figure 4 illustrates the creation of two similar record-level Bloom filters.

HLSH utilizes L independent hash tables. Each hash table, denoted by T_l where $l = 1, \ldots, L$, consists of key-bucket pairs where a bucket hosts a linked list which is aimed at grouping similar Bloom filter pairs. Each hash table has been assigned a composite hash function g_l which consists of a fixed number K of base hash functions. A base hash function applied to a Bloom filter returns the value of its j-th position where $j \in \{0, \ldots, S-1\}$ chosen uniformly at random.

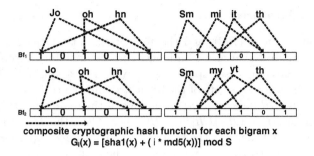

Fig. 4. Creating two similar Bloom filters from similar strings.

We assume a pair of Bloom filters of distance less than or equal to a predefined threshold as similar. The smaller the Hamming distance of a Bloom filter pair is, the higher the probability for a g_l to produce the same result. The result of a g_l, which constitutes the blocking key and is applied to a Bloom filter, specifies into which bucket of some T_l, this Bloom filter will be stored. The intuition behind HLSH is the random choices of bits, which are performed redundantly, and result in grouping similar Bloom filters in at least one T_l. During the matching step, we scan the buckets of each T_l and formulate pairs.

The optimal number of the T_l's that should be utilized is:

$$L = \lceil \frac{\ln(\delta)}{\ln(1 - p^K)} \rceil,$$

where p denotes the probability of a base hash function of producing the same result by hashing two similar Bloom filters. The similarity of a Bloom filter pair is user-defined and denoted by θ. By using this structure, each similar Bloom filter pair will be returned with high probability $1 - \delta$, as δ is usually set to a small value, say $\delta = 0.1$. A method, which relies on random sampling of Bloom filter pairs, for choosing the optimal value of K is presented in [18].

5   Evaluation

We evaluate the above-mentioned methods in terms of (a) the accuracy in finding the truly matched record pairs, (b) the efficiency in reducing the number of candidate pairs, and (c) the execution time. We use semi-synthetic data sets, denoted by A and B, with size of $1,000,000$ records each, extracted from the NCVR [23] list. Each record contains as fields the *LastName*, *FirstName*, *Address*, and *Town*. Insert, edit, delete, and transpose operations, chosen in random order, are used to perturb the values of each field of certain marked records from A. Eventually, those perturbed records are placed in data set B. For the experiments, we used a simple PC with an Intel i5-2400 and $16\,GB$ RAM. The software components were developed using the Java programming language (JDK 1.7)

5.1   Selected Measures

The Pairs Completeness (PC), and the Reduction Ratio (RR) metrics [4] are employed to evaluate the accuracy in identifying the matching record pairs and the reduction of the comparison space, respectively. PC is equal to $|\mathcal{M} \cap M|/|M|$, where $\mathcal{M}$ and M denote the sets of the identified and the truly matching record pairs, respectively. RR indicates the percentage of the reduction of the comparison space given by $RR = 1.0 - |CR|/|A \times B|$, where CR is the set of the candidate record pairs and $A \times B$ denotes the comparison space. Each experiment was run 50 times and we plotted the average values in the figures shown below.

5.2   Configuration Parameters

In order to achieve higher accuracy rates in EUC, we (a) set 30 dimensions for each field, and (b) turned off both the greedy re-sampling and the distance approximation heuristic, both illustrated in [26]. For the LSH-based mechanism, we set $K = 5$, and thresholds to 4.5 and 8 for each perturbation scheme, respectively. We experimented with several values for L and we chose $L = 30$ and $L = 140$ which achieve a good balance between efficiency and accuracy. In AHC, we set the *LastName* field as the blocking field, the privacy parameter ϵ to 0.5, and the number of clusters to 500. We did not apply any negative noise in order not to suppress any records from those clusters. For HG, we categorized the records into the educational hierarchies, which were assigned randomly to records, as illustrated in [13]. In TPB, we used the configuration parameters proposed by the authors in [30], which are $k = 100$, $w = 2$, and $n_e = 50\%$. For HLSH, we set the distance threshold $\theta = 200$, the size of the record-level Bloom filters $S = 2,000$, $K = 30$, and $\delta = 0.1$. Using these parameters, HLSH generated $L = 54$ blocking groups.

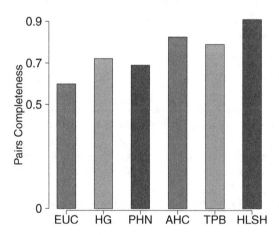

Fig. 5. Measuring the Pairs Completeness (accuracy) rates.

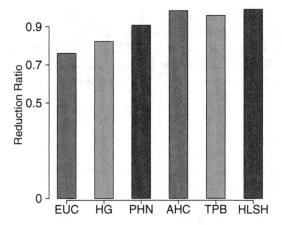

Fig. 6. Measuring the Reduction Ratio.

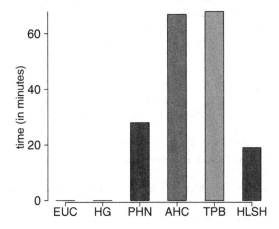

Fig. 7. Measuring the time consumed.

5.3 Comparative Results

Figure 5 illustrates the PC rates of each method and it can be clearly seen that HLSH, AHC, and TPB achieve the highest scores. However, we have to note that the performance of AHC and TPB is highly dependent on the choice of the reference values. We tested several sets of reference values and achieved high PC rates only when those sets were supersets of the field values. Conversely, if those sets were not supersets of the field values, the PC rates dropped considerably below 70%. HG and PHN exhibited stable performance, while EUC had large deviations from its mean rate mainly due to the embedding method and the random formulation of the reference set of strings.

The reduction of the comparison space, as measured by the RR, is shown in Fig. 6. HLSH, AHC, TPB, and PHN exhibit comparable performance reaching almost *98%* reduction. Finally, Fig. 7 demonstrates the running time consumed, where HLSH outperforms clearly all the other methods. Also, we notice that the rates of PHN are quite close to those of HLSH. To be fair though, AHC and TPB perform SMC computations, which are reliable but are computationally expensive, during the matching step. EUC and HG crashed the system throwing the *'Out of Memory'* error.

6   Conclusions

Linking large collections of records by simultaneously protecting their privacy has arisen recently as an intriguing problem in the core of the domain known as Privacy-Preserving Record Linkage. In this paper we compare the HLSH method, which relies on the Locality-Sensitive Hashing technique and the Bloom filter-based encoding method, with five state-of the-art private blocking methods. HLSH outperformed these methods in terms of the accuracy of the results as well as the running time required.

References

1. Aggarwal, C., Yu, P.: The IGrid index: reversing the dimensionality curse for similarity indexing in high dimensional space. In: SIGKDD, pp. 119–129 (2000)
2. Al-Lawati, A., Lee, D., McDaniel, P.: Blocking-aware private record linkage. In: IQIS, pp. 59–68 (2005)
3. Christen, P.: Data Matching - Concepts and Techniques for Record Linkage, Entity Resolution, and Duplicate Detection. Data-Centric Systems and Applications. Springer, Heidelberg (2012)
4. Christen, P.: A survey of indexing techniques for scalable record linkage and deduplication. TKDE **24**(9), 1537–1555 (2012)
5. Cohen, W., Richman, J.: Learning to match and cluster large high-dimensional datasets for data integration. In: SIGKDD, pp. 475–480 (2002)
6. Datar, M., Immorlica, N., Indyk, P., Mirrokni, V.: Locality-sensitive hashing scheme based on p-stable distributions. In: Symposium on Computational Geometry, pp. 253–262 (2004)
7. Durham, E.: A Framework For Accurate Efficient Private Record Linkage. Ph.D. thesis, Vanderbilt Univ., US (2012)
8. Dwork, C.: Differential privacy. In: Bugliesi, M., Preneel, B., Sassone, V., Wegener, I. (eds.) ICALP 2006. LNCS, vol. 4052, pp. 1–12. Springer, Heidelberg (2006)
9. Gionis, A., Indyk, P., Motwani, R.: Similarity search in high dimensions via hashing. In: VLDB, pp. 518–529 (1999)
10. Goodman, J., O'Rourke, J., Indyk, P.: Handbook of Discrete and Computational Geometry. CRC, Boca Raton (2004)
11. Hall, R., Fienberg, S.E.: Privacy-preserving record linkage. In: Domingo-Ferrer, J., Magkos, E. (eds.) PSD 2010. LNCS, vol. 6344, pp. 269–283. Springer, Heidelberg (2010)
12. Hernandez, M., Stolfo, S.: Real world data is dirty: data cleansing and the merge/purge problem. DMKD **2**(1), 9–37 (1988)

13. Inan, A., Kantarcioglou, M., Bertino, E., Scannapieco, M.: A hybrid approach to private record linkage. In: ICDE, pp. 496–505 (2008)
14. Inan, A., Kantarcioglu, M., Ghinita, G., Bertino, E.: Private record matching using differential privacy. In: EDBT, pp. 123–134 (2010)
15. Jin, L., Li, C., Mehrotra, S.: Efficient record linkage in large datasets. In: DASFAA, pp. 137–146 (2003)
16. Karakasidis, A., Verykios, V.: Privacy preserving record linkage using phonetic codes. In: BCI, pp. 101–106. IEEE (2009)
17. Karakasidis, A., Verykios, V.: A sorted neighborhood approach to multidimensional privacy preserving blocking. In: ICDM Workshops, pp. 937–944 (2012)
18. Karapiperis, D., Verykios, V.: A distributed near-optimal LSH-based framework for privacy-preserving record linkage. COMSIS 11(2), 745–763 (2014)
19. Karapiperis, D., Verykios, V.: A distributed framework for scaling up LSH-based computations in privacy preserving record linkage. In: BCI, pp. 102–109. ACM (2013)
20. Karapiperis, D., Verykios, V.: An LSH-based blocking approach with a homomorphic matching technique for privacy-preserving record linkage. TKDE 27(4), 909–921 (2015)
21. Kim, H., Lee, D.: Fast iterative hashed record linkage for large-scale data collections. In: EDBT, pp. 525–536 (2010)
22. Kuzu, M., Kantarcioglu, M., Inan, A., Bertino, E., Durham, E., Malin, B.: Efficient privacy-aware record integration. In: EDBT, pp. 167–178 (2013)
23. NCVR: North Carolina voter registration database. ftp://www.app.sboe.state.nc. us/data
24. Paillier, P.: Public-key cryptosystems based on composite degree residuosity classes. In: Stern, J. (ed.) EUROCRYPT 1999. LNCS, vol. 1592, pp. 223–238. Springer, Heidelberg (1999)
25. Rivest, R.: Chaffing and winnowing: Confidentiality without encryption. MIT Internal paper (2011)
26. Scannapieco, M., Figotin, I., Bertino, E., Elmagarmid, A.: Privacy preserving schema and data matching. In: SIGMOD, pp. 653–664 (2007)
27. Schnell, R., Bachteler, T., Reiher, J.: Privacy-preserving record linkage using bloom filters. BMC Med. Inf. Decis. Mak. 9(41), 1–11 (2009)
28. Sweeney, L.: k-anonymity: a model for protecting privacy. Uncertainty Fuzziness Knowl. Based Syst. 10(5), 557–570 (2002)
29. Vatsalan, D., Christen, P., Verykios, V.: An efficient two-party protocol for approximate matching in private record linkage. In: AUSDM, pp. 125–136 (2011)
30. Vatsalan, D., Christen, P., Verykios, V.: Efficient two-party private blocking based on sorted nearest neighborhood clustering. In: CIKM, pp. 1949–1958 (2013)
31. Vatsalan, D., Christen, P., Verykios, V.: A taxonomy of privacy-preserving record linkage techniques. Inf. Sys. 38(6), 946–969 (2013)
32. Weber, R., Schek, H., Blott, S.: A quantitative analysis and performance study for similarity search methods in high dimensional spaces. In: VLDB, pp. 194–205 (1998)
33. Yakout, M., Atallah, M., Elmagarmid, A.: Efficient private record linkage. In: ICDE, pp. 1283–1286 (2009)

Regular Contributions

Secret Shared Random Access Machine

Shlomi Dolev and Yin Li[✉]

Department of Computer Science,
Ben-Gurion University of the Negev, Beersheba, Israel
dolev@cs.bgu.ac.il, yunfeiyangli@gmail.com

Abstract. The computations over RAM are preferred over computations with circuits or Turing machines. Secure and private RAM executions become more and more important in the scope avoiding information leakage when executing programs over a single computer as well as over the clouds. In this paper, we propose a distributed scheme for evaluating RAM programs without revealing any information on the computation including the program, the data and the result. We use the Shamir secret sharing to share all the program instructions and private string matching technique to ensure the correct instruction execution. We stress that our scheme obtains information theoretic security and does not rely on any computational hardness assumptions, therefore, gaining indefinite private and secure RAM execution of perfectly unrevealed programs.

Keywords: Shamir secret sharing · Random access machine · Information theoretic secure

1 Introduction

Cloud computing provides cost-efficient and flexible shared infrastructure and computational services on demand for various customers who need to store and operate on a huge amount of data. Until now, there are various services providers such as Amazon [1] and Google [13] offering platforms, software, and storage outsourcing applications. Much attention has been paid to them due to the potential benefits and business opportunities that clouds could bring [9].

However, cloud computing also introduces security and privacy risks for the clients. For example, some of the cloud providers are not perfectly reliable and are vulnerable to network attacks and data leakage. Furthermore, even a single computer with the same cloud organization is untrustworthy. There are possible attacks on a single computer during which information is copied from the bus of the computer and sent to an adversary.

Several techniques are applied to address data storage privacy [18–20, 26] and security computation on clouds [17, 29]. Among these studies, security

---

S. Dolev—Partially supported by Kamin grant of the Israeli economy ministry, and the Rita Altura Trust Chair in Computer Sciences.

Y. Li—The author would like to acknowledge the Lynne and William Frankel Center as it supports students travel for presenting their works.

© Springer International Publishing Switzerland 2016
I. Karydis et al. (Eds.): ALGOCLOUD 2015, LNCS 9511, pp. 19–34, 2016.
DOI: 10.1007/978-3-319-29919-8_2

in evaluating random access machine (RAM) program is an important task [2,23], since many modern algorithms are operating on the von Neumann RAM architecture. Until now, there are mainly two ways, the first is to convert a RAM program into circuits and the second is to use oblivious RAM, introduced by Goldreich and Ostrovsky [19]. Oblivious RAM schemes are preferred as there is no need to convert the program into a binary circuit which leads to a huge blowup in program size and its running time.

Even though the propositions for secure RAM evaluation can address various privacy challenges including two-party [22,23], multiparty [5,10] or large-scale computation [6] against semi-honest or malicious adversaries, they all assume that the processors used by clouds are trustworthy. Thus, in these proposals, the CPU has to decrypt the input data before processing and then encrypt the output data again. In fact, an adversary can introduce a special hardware Trojan [28] designed to disable or destroy a system in the future, or leak confidential information. Similar attack has already been demonstrated in [3], where a specially designed Trojan in the CPU revealed sensitive information to the adversary.

Threat Model. We assume that there is a client that wants to run a program on the clouds. But the client does not want to reveal any information about both the program and the data. The adversary, has deployed the untrusted hardware to the clouds. That is to say, the adversary can listen to the bus, may extract information on the internal activity of the processor. All the clouds are not necessarily semi-honest.

Unfortunately, none of the above protocols can avoid information leakage under such threat model. Thus, one may wish to execute an encrypted program on encrypted data without decrypting neither the program nor the data. A straightforward approach is to execute the encrypted instructions in the clouds processors directly. Fully homomorphic encryption [14,15] (FHE) is a way to achieve this goal. Several schemes are proposed to implemented secret program execution over FHE (e.g., [7,8,31]). However, the main problem is that the proposed schemes have high overhead of computation [16] which make FHE more theoretical result than practical. Moreover, Gentry's scheme and later FHE schemes relied on the hardness assumptions such that of the ideal lattices, which are only computationally secure, rather than key-less information theoretical secure.

Our Contribution. In this paper, an alternative architecture is proposed with security and privacy that are based on theoretically security promises. The main technique is a combination of Shamir Secret Sharing [25] and the recently proposed Accumulating Automata [12].

Secret sharing is used to utilize perfect privacy of the client's program and processor states and secret string matching [12] is used to facilitate instruction execution. We note that the modern instruction set, for example, CISC and RISC, originally designed for efficiency and performance [21], are too complicated when there is a need to hide their nature of operation and the sequence of operations they form. Thus we apply One Instruction Set Computer (OISC) to our model. We simulate the OISC instruction *subtract and branch if less than or equal to zero* (Subleq) that is proven complete and for which there exists a compiler from high-level programming languages to Subleq [24]. As a result, our scheme has the following significant characteristics

- *Information theoretic security.* We use Shamir secret sharing which could provide information theoretic security for clients. In our scheme, the user's program is secret shared and run on independent machines and clouds. Each cloud only needs to perform computation without communicating with other clouds. Moreover, note that we use the instruction Subleq proven to be complete in terms of Turing-complete computation. Thus, our model can execute any RAM programs privately and securely.
- *Program protection.* During the whole execution of the program, for every instruction, the processors have to "touch" all the instructions in the memory. Moreover, for every data access, the processors also have to access all of the data items. The execution mode and access pattern make the client program "oblivious" to the clouds, thus ensuring both data and program privacy. Still, the operations can be delegated by the users to powerful machines in the clouds, which perform these linear access to all items for executing operations without revealing their nature and sequence.
- *Error correcting.* Notice that the clients run their programs in E independent machines/clouds. According to the conclusion of Ben-Or et al. [4], as long as less than one-third of clouds are malicious (do not follow the protocol possibly returning wrong results), the client can detect their actions by reconstructing the final result using Lagrange interpolation.

The rest of the paper is organized as follows: in Sect. 2, we briefly introduce the settings used in our paper. Section 3 describes the basic primitives we use in our construction. Explicit application and its security analysis are given in Sect. 4. Finally, conclusions are drawn in Sect. 5.

2   Preliminary

In this section, we briefly introduce the basic ingredients used in the sequel.

Shamir Secret Sharing. Shamir secret sharing (SSS) is an information theoretic secure protocol, which allows a dealer to secret share a values s among E players. There is a threshold δ for the scheme, such that, the knowledge of δ or fewer player secrets make the adversary learn no information about s, but if more than δ players communicate their shares to each other, they can easily recover the secret.

Distribution: The dealer picks a random polynomial $f \in \mathbb{F}_p[x]$ of degree $\delta < E$ such that $f(0) = s \in \mathbb{F}_p$. The dealer also chooses E arbitrary non-zero indices $\alpha_1, \cdots, \alpha_E$, computes $f(\alpha_i)$ for $1 \leq i \leq E$ and send $(\alpha_i, f(\alpha_i))$ to each corresponding players.

Reconstruction: Any $\delta + 1$ players can reconstruct the polynomial f by applying Lagrange interpolation to the tuples $(\alpha_i, f(\alpha_i))$. They recover the secret by computing $f(0) \bmod p = s$.

Shamir secret sharing is additively homomorphic but is not multiplicatively homomorphic. Namely, if we want to perform multiplication using Shamir secret shares, a special "degree reduction step" is required. We will discuss this problem more explicitly in the following section.

Private String Matching. Recently, Dolev et al. proposed a secret string matching algorithm using Accumulating Automata [12]. The algorithm runs on several cloud servers. The strings to be compared are originally secret shared using Shamir secret sharing and therefore stay unknown to the processing servers. Note that the comparison of two strings represented in secret shares is different from the comparison of strings in a plaintext format, as each participant cannot judge the compare result independently.

Unary representation: The authors of [12] demonstrated their scheme over unary letter representation, where each letter is represented by a binary number with hamming weight 1. For example, letter a–z are expressed by the binary strings: $a = [100\cdots00], b = [010\cdots00], c = [001\cdots00], \cdots, z = [000\cdots01]$ with each representation consists of 26 bits. We use the expression $S = \sum_{i=0}^{r} u_i \times v_i$, to compare two letters, where $[u_0 u_1 \cdots u_r]$ and $[v_0 v_1 \cdots v_r]$ are two unary representations. It is clear that whenever the two representations are identical, S is equal to 1, otherwise S is equal to 0. Assume that each cloud has the secret shares of these two representations, i.e., $(\alpha, f_i(\alpha))$ and $(\alpha, g_i(\alpha))$, where $f_i(0) = u_i$ and $g_i(0) = v_i$. Similarly, it can compute the following equation to identify whether the two letters are identical:

$$\sum_{i=1}^{r}(f_i(\alpha) \times g_i(\alpha)). \tag{1}$$

We have following lemma.

Lemma 1. *If the two letters are identical, then the result of Eq. (1) is the secret share of 1, otherwise the result of this equation is a secret share of 0.*

Proof. Note that u_i, v_i are the secret bit and would be either 1 or 0. Let $f_i'(\alpha)$ and $g_i'(\alpha)$ denote the evaluation of $f(x)$ and $g(x)$ at point α without the constant term u_i, v_i, respectively. We can see

$$
\begin{aligned}
&f_i(\alpha) \times g_i(\alpha) \\
&= (f_i'(\alpha) + u_i) \times (g_i'(\alpha) + v_i) \\
&= f_i'(\alpha)g_i'(\alpha) + u_i g_i'(\alpha) + v_i f_i'(\alpha) + u_i v_i \\
&= F(\alpha) + u_i v_i,
\end{aligned}
$$

where $F(\alpha) = f_i'(\alpha)g_i'(\alpha) + u_i g_i'(\alpha) + v_i f_i'(\alpha)$. Therefore, $f_i(\alpha) \times g_i(\alpha)$ can be seen as a secret share of $u_i v_i$. It is clear that only when $u_i = v_i = 1$, $f_i(\alpha) \times g_i(\alpha)$ is a secret share of 1, and otherwise it is a secret share of 0. Note that the hamming weight of unary representation is only 1, one can directly find the finial summation is at most 1 which conclude the result.

Based on this observation, it is easy to compare a string using Accumulating Automata, which is a type of finite automata. Only when the string letters are exactly the same, the last node will be set to 1, otherwise this node will stay 0. One can reconstruct the values of this node to identify whether the string matching is successful or not.

Binary representation: The main drawback of unary representation is that it has too many redundant bits. For example if we want to represent the numbers 1 to 1000, we have to use 1000 bits. An alternative method is to use binary representation.

Assume that there are two letters represented as $[u_0 u_1 \cdots u_r]_2$ and $[v_0 v_1 \cdots v_r]_2$, where $u_i, v_i \in \{0, 1\}$. We compare these letters using the Algorithm 1.

As a simple example, we consider two binary strings $[1010]_2$ and $[1101]_2$. According to previous description, we perform the following computations:

Algorithm 1. Secret comparison using binary representation

1: **for** $i = 1$ to r **do**
2:     $s_i = [u_i - v_i]^2$
3: **end for**
4: $S = 0$
5: **for** $i = 1$ to r **do**
6:     $S = S + s_i - S \times s_i$
7: **end for**
8: return $1 - S$

– Bitwise subtraction,
  $[1, 0, 1, 0] - [1, 1, 0, 1] = [1 - 1, 0 - 1, 1 - 0, 0 - 1] = [0, -1, 1, -1]$;
– Bitwise squaring,
  $[0^2, (-1)^2, 1^2, (-1)^2] = [0, 1, 1, 1]$;
– Bitwise OR, $S = 0|1|1|1 = 1$;[1]

It is easy to check that if the two strings are equal, S is equal to 0 and otherwise to 1. In this example, the value of S is 1. In order to return the same value as the unary representation, we prefer to return $1 - S$ rather than S. Note that we only use the subtraction/addition and multiplication in the above algorithm, similarly to the unary case, these operations can also be implemented using Shamir secret sharing. However, compared with unary representation, it requires either more participants or (more) degree reduction operations.

One Instruction Computer Set. OISC is an abstract machine that uses only one instruction. It is proven that OISC is capable of being a universal computer in the same manner as traditional computers with multiple instructions [24]. This indicates that one instruction set computers are very powerful despite the simplicity of the design, and can achieve high throughput under certain configurations.

Since there is only one instruction in the system, it needs no identification to determine which instruction to execute. Thus, we only need to design the implementation of one instruction. Actually, there are several options for choosing the OISC instruction, such as *subtract and branch if not equal to zero* (SBNZ), *subtract and branch if less than or equal to zero* (Subleq), *add and branch unless positive* (Addleq). Among these instructions, Subleq is the most commonly used. Nowadays, there are Subleq compiler and Subleq-based processor [27] which

---

[1] One can check that Step 6 in Algorithm 1 is equivalent to the bitwise OR operation.

make Subleq a practical and efficient choice. Therefore, in this paper, we focus
on how to simulate Subleq privately and secretly. Comparing the values of two
memory words that are represented by secret shares, is hard to implement, hence
we secret share the words bit by bit, perform the arithmetic over secret shared
bits and then branch according to the sign bit of the result. This leads to a novel
scheme for executing secret shared Subleq (SSS-Subleq) programs. The details
are presented in Sect. 3.

3   SSS-Subleq Programs and Their Execution

Since our architecture is built on Subleq, for any client programs written by
high-level languages, it needs to be compiled into Subleq codes at first [27].
Then the client executes the set of Subleqs over the system. In the following,
we will investigate the implementation details of Subleq using Shamir secret
sharing.

The SSS-Subleq Format and Architecture Overview. According to the
definition of Subleq, it has three parameters A, B, C where the contents at
address B are subtracted from the contents at address A, and the result is
stored at address B, and then, if the result is not greater than 0, the execution
jumps to the memory address C, otherwise it continues to the next instruction
in the sequence. The pseudo code is given in the procedure $\text{Subleq}(A, B, C)$.
Here, the PC (program counter) is a pointer that indicates the address of next
instruction.

Note that the Subleq contains
some important operations: load,
store, subtraction and conditional
branch. Thus, in order to execute Subleq using Shamir secret sharing, we
have to simulate the following operations using secret shares:

Procedure. $\text{Subleq}(A, B, C)$
1: $\text{Mem}[B] = \text{Mem}[B] - \text{Mem}[A]$
2: **if** $\text{Mem}[B] \leq 0$ **then**
3:      **goto** C
4: **else**
5:      **goto** $PC + 1$
6: **end if**

- $\text{LOAD}(H)$: Load the instruction in address H to the processor.
- $\text{JUMP}(C)$: Transfers control to index C, implement the branching operation.
- $\text{READ}(X)$: Read the data at address X.
- $\text{WRITE}(X, Y)$: Write the data Y in address X.

Please note that the operation goto $PC + 1$ and goto C can be implemented
by the operation JUMP with different parameters. Among all these operations,
a critical problem is how to find the right address secretly. Fortunately, secret
string matching allows us to implement these operations without revealing any
information. According to the description in Sect. 2, we use unary representation
to represent the addresses including memory addresses and instruction indices
where each bit of the unary representation is encoded as a secret shared value.
The format of the SSS-Subleq instruction has five parts which are shown in
Fig. 1.

The first block stores the instruction index number which is equivalent to the instruction address, the second and third blocks store the operand addresses and the fourth to fifth blocks

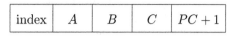

Fig. 1. Format of SSS-Subleq

store the branch index C and the index of next instruction, respectively.

Besides the former operations, there is a need to implement the subtraction between two operands and determine the next instruction address according to the subtraction result. Therefore, we choose to represent every operand as a signed number. In order to perform subtraction in an easy way, we use two's complement representation where subtraction can be transformed into addition. The most significant bit (MSB) is the sign bit. Analogous with the address, each bit of the operands is secret shared. The outline of our RAM architecture is presented Fig. 2. In our architecture, we use a modified Harvard architecture which not only physically separates storage and signal pathways for instructions and data, but also separates the read-only and read/write part of data. Note that since Shamir secret sharing is not multiplicatively homomorphic, degree reduction is needed after several multiplications. This special structure allows us to implement read and write operations in relatively efficient manner. In particular, the degree of the polynomials used for the read-only part (possibly big-data corpus) is unchanged throughout the execution(s).

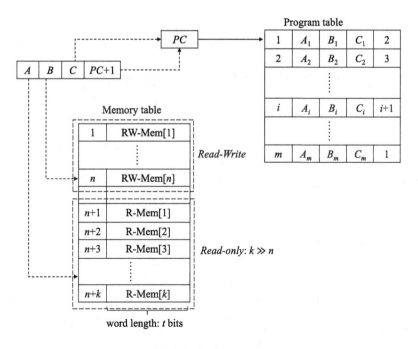

Fig. 2. Architecture

Table 1. The parameters of a program

Parameter	Description
m	The number of instructions of the user program
n	The number of data items that can be accessed for read and write
k	The number of data items that can be accessed for read only
t	The bit length of the data stored in the memory

The parameters of our architecture are presented in Table 1. Here, we assume that the client program reads a large number of data items compared with the data items the program writes to, thus we have $k \gg n$. In the following, we will show how to simulate the four basic operations using the Accumulating Automata technique.

Operation Details. We start describing the implementation of a function called: $compare(U, V, r)$, where U and V are secret shares of the unary address consisting of r elements. For example, let $U = u_1, u_2, \cdots, u_r$, $V = v_1, v_2, \cdots, v_r$ denote the secret shares of two such parameters, we compute

$$compare(U, V, r) = \sum_{i=1}^{r}(u_i \times v_i) \tag{2}$$

According to Sect. 2, the above expression testifies whether U, V are identical or not. It is easy to check that the result of $compare(U, V, r)$ is a secret share of 1 if $U = V$, and otherwise, if $U \neq V$, is 0.

Now we describe the details of the four operations:

Description of LOAD: The initial values of S_i are set to 0, and the symbol $\|$ means concatenation of all values from S_1 to S_4. H represents the secret shares of the instruction address which we want to load and η_i represents secret shares of the i-th instruction address. It is clear that the value returned is the right instruction we want to load.

Procedure. LOAD(H)

1: **for** $i = 1$ to m **do**
2:      $Num_i \leftarrow compare(H, \eta_i, m)$
3:      $S_1 \leftarrow S_1 + Num_i \times A_i$
4:      $S_2 \leftarrow S_2 + Num_i \times B_i$
5:      $S_3 \leftarrow S_3 + Num_i \times C_i$
6:      $S_4 \leftarrow S_4 + Num_i \times (PC_i + 1)$
7: **end for**
8: return $S_1\|S_2\|S_3\|S_4$

Description of READ: According to Fig. 2, the format of memory table consists of two parts: the address number ϵ_i and data θ_i. Analogous to the corresponding analysis for the LOAD operation, we can easily check that S is the data whose index number is equal to X.

Procedure. READ(X)

1: **for** $i = 1$ to $n + k$ **do**
2:      $Num_i \leftarrow compare(X, \epsilon_i, n + k)$
3:      $S \leftarrow S + Num_i \times \theta_i$
4: **end for**
5: return S

Description of WRITE: The operation implements writing the data Y in the address X using secret shares. Note that only when ϵ_i equals X, the Num_i is the secret shares of 1, and then the data Y can substitute the former data item, otherwise the data will not be changed.

Procedure. WRITE(X, Y)

1: **for** $i = 1$ to n **do**
2:     $Num_i \leftarrow compare(X, \epsilon_i, n + k)$
3:     $\theta_i \leftarrow \theta_i + Num_i \times (Y - \theta_i)$
4: **end for**

Description of JUMP: The operation JUMP is nearly the same as the operation LOAD. If the program needs to execute the C-th instruction in the pro-

---

Procedure. JUMP(C)

1: $PC \leftarrow C$
2: LOAD(PC)

---

gram table, it just assigns the last part of current instruction to the PC. Then the program will "jump" to the destination.

Implementation of SSS-Subleq. We then investigate the conditional branch that required in Subleq in secret shares form. It is difficult to compare two numbers directly since all the numbers are secret shared and the clouds never know the secrets. Here, we use two's complement to represent the operands and using the sign bit to implement the comparison. In two's complement, the sign bit of positive integer is 0 and negative integer is 1. Therefore, when implementing SUBLEQ(A, B, C), we can use the sign bit of $Mem[B] - Mem[A]$ to (blindly) decide whether to branch or not. The only problem is that the integer 0, for which the sign bit in its representation is also 0, while it should imply branching. This problem can be fixed by a slight modification: using the sign bit of $Mem[B] - Mem[A] - 1$ instead of sign bit of $Mem[B] - Mem[A]$. Moreover, we will show that this sign bit can be obtained during the computation of $Mem[B] - Mem[A]$ in the following paragraphs.

Two's Complement Subtraction. The advantage of using two's complement is the elimination of examining the signs of the operands to determine if addition or subtraction is needed. Therefore, to compute subtraction $\beta - \alpha$, it only need to perform following steps:

– Convert α: Invert every bit of α and add one, denoted by $\bar{\alpha} + 1$.
– Addition: Perform binary addition and discard any overflowing bit, denoted by $\beta + \bar{\alpha} + 1$.

Note that we also need the sign bit of $\beta - \alpha - 1$. As described above, using two's complement representation, the subtraction $\beta - \alpha$ is converted to $\beta + \bar{\alpha} + 1$. Similarly, the subtraction $\beta - \alpha - 1$ is implemented as

$$\underline{\beta - \alpha - 1} = \underline{\beta + \bar{\alpha} + 1} - 1 = \beta + \bar{\alpha}.$$

The similarity allows us to implement these two subtractions together.

The algorithm for two's complement subtraction using Shamir secret sharing is given in Algorithm 2. According to previous description in Sect. 2, we know

---

Algorithm 2. The two's complement subtraction using Shamir secret sharing

---

1: **procedure** SSS-SUB(A, B)
2:     **Input:** $A = [a_{t-1}a_{t-2} \cdots a_1 a_0], B = [b_{t-1}b_{t-2} \cdots b_1 b_0]$ where a_i, b_i are secret shares of bits of two's complement represented number.
3:     **Output:** $R = [r_{t-1}r_{t-2} \cdots r_1 r_0]$ where $R = B - A$, and the sign bit of $B - A - 1$
4:     $a_0 = 1 - a_0$ ▷ Invert of the least significant bit
5:     $carry[0] = a_0 \cdot b_0$
6:     $r_0 = a_0 + b_0 - 2 \cdot carry[0]$ ▷ Addition of the least significant bit
7:     **for** $i = 1$ to $t - 1$ **do**
8:        $a_i = 1 - a_i$ ▷ invert each bit $A \rightarrow \bar{A}$
9:        $r_i = a_i + b_i - 2a_i b_i$
10:       $carry[i] = a_i b_i + carry[i - 1] \cdot r_i$ ▷ The carry bit
11:       $r_i = r_i + carry[i - 1] - 2 \cdot carry[i - 1] \cdot r_i$ ▷ The result bit
12:     **end for**
13:     $sign = r_{t-1}$ ▷ The sign bit of $B - A - 1$, used for branch
14:     $carry[0] = r_0$ ▷ Add 1 to the result obtain $B - A$
15:     $r_0 = 1 - r_0$
16:     **for** $i = 1$ to $t - 1$ **do**
17:       $carry[i] = r_i \cdot carry[i - 1]$
18:       $r_i = r_i + carry[i - 1] - 2 \cdot carry[i]$
19:     **end for**
20:     **return** $(R \| sign)$
21: **end procedure**

Algorithm 3. The Shamir secret sharing based Subleq

1: **procedure** SSS-SUBLEQ(A, B, C)
2:     $R_1 \leftarrow \text{READ}(A)$
3:     $R_2 \leftarrow \text{READ}(B)$
4:     $R \| Num = \text{SSS-SUB}(R_1, R_2)$
5:     $\text{WRITE}(B, R)$
6:     $\text{JUMP}(Num \cdot C + (1 - Num) \cdot (PC + 1))$
7: **end procedure**

---

the multiplications and additions/subtractions of the shares correspond to those of the secrets. Thus one can easily check that Algorithm 2 implements the two's complement subtraction.

Therefore, Subleq can be implemented with secret shares by Algorithm 3. In step 6, we can check that if the value represented by R_2 is less than R_1, then $Num = 1, PC = C$, else $Num = 0, PC = PC + 1$. Therefore, this expression implement the conditional branch of Subleq.

Degree Reduction. The main bottleneck of our scheme is the multiplication with shares used in the basic operations, as the Shamir secret sharing is not multiplication homomorphic. Note that multiplying two polynomials gives a polynomial with a degree that is equal to the sum of the degrees of the source polynomials. We observe that the READ, JUMP and LOAD increase the polynomial degrees related to each secret shared bit stored in the registers, the subtraction and WRITE increase the degrees related to the data items stored in the memory. Therefore, we have to process the degree reduction for these data items at a certain time. In [11], Dolev et al. proposed a method for reducing the polynomial degree without revealing the original secret. In our model, we define a *reducer* that is in charge of reducing the polynomial degrees and a *randomizer* in charge of generating random polynomials for all the participants. Note that the codes of the *reducer* and the *randomizer* should be executed independently in order to protect the secret s, but either of them can be executed by the dealer machine. The polynomial degree reduction algorithm appears in Algorithm 4.[2]

Algorithm 4. Polynomial degree reduction for secret shares

1: **procedure** DECREASE($P(x), d, d^*$)
2:     Let $u_1, \cdots, u_E$ be E participants, D be the *randomizer* and R be the *reducer*.
3:     Let $P(x) \in \mathbb{F}_p[x]$ of degree d is the polynomial for secret s.
4:     D randomly selects polynomial $f(x)$ of degree d and $g(x)$ of degree d^*, where $f(x)$ and $g(x)$ have the same constant term.
5:     **for** $i = 1$ to E **do**
6:         D sends $(f(u_i), g(u_i))$ to u_i.
7:         u_i computes $P(u_i) + f(u_i)$ and sends it to R.
8:     **end for**
9:     R interpolates and computes a polynomial $Q(x) = P(x) + f(x)$.
10:    **for** $i = 1$ to E **do**
11:        R sends to u_i the coefficients q_j of $Q(x)$ with degree more than d^*.
12:        u_i computes $S = P(u_i) + f(u_i) - \sum_{j=d^*+1}^{d} q_j u^j - g(u_i)$.
13:        **return** S.
14:    **end for**
15: **end procedure**

Different from the original algorithm presented in [11], we use the random polynomials $f(x)$ of degree d instead of d^*. It is clear that adding $f(x)$ to $P(x)$ can hide all the coefficients of $P(x)$ which prevent the *reducer* from obtaining any information about the secret s. We also use another random polynomial $g(x)$ of degree d^*, where the constant term of $f(x)$ and $g(x)$ are identical. In the end of Algorithm 4, each cloud subtracts $g(u_i)$ from the result which will keep the secret s unchanged To protect the secrets, for every degree reduction, the random polynomial $f(x), g(x)$ should be updated. In practical implementation, the dealer (with no *randomizer*) can secret share these polynomials to the clouds in advance or let clouds interact with the *randomizer*, thus supplies on-line these $f(x)$ and $g(x)$ pairs upon requests and the degree reduction process is performed with no involvement of the dealer during the execution.

In our proposed architecture, the read/write memory is separated from the read-only memory. This design is more convenient for degree reduction compared with the classic architecture. Compared with the whole memory space,

---

[2] The original algorithm is designed for bivariate polynomial, we modified it accordingly.

Algorithm 5. The SSS-Subleq plus degree reduction

1: **procedure** SSS-SUBLEQ-DR(A, B, C)
2:      $Decrease(A\|B\|C\|PC + 1, 3\ell, \ell)$
3:      $R_1 \leftarrow \text{READ}(A)$
4:      $R_2 \leftarrow \text{READ}(B)$
5:      $R\|Num = \text{SSS-SUB}(R_1, R_2)$
6:      $\text{DECREASE}(R\|Num, *, \ell)$
7:      $\text{WRITE}(B, R)$
8:      $\text{JUMP}(Num \cdot C + (1 - Num) \cdot (PC + 1))$
9: **end procedure**

---

the read/write registers are very small, thus, the number of items for which we need to reduce the degree is relatively small. Assume that both the addresses and data items are secret shared using the polynomials of the same degree ℓ, plus degree reduction step, the Subleq can be implemented as in Algorithm 5. In step 6, we use $*$ instead of the exact degree parameter, as each secret shared bit of R has different polynomial degree.

4 Applications

In our model, assume that a client wants to outsource a program in clouds and the program is compiled into Subleq-based program. The address is encoded using unary representation and the data item is encoded using two's complement representation. The dealer picks random polynomials of degree ℓ to share every bit of the address and data. Then the dealer sends the secret shared program to E clouds. The integer E should be greater than the highest polynomial degree generated during Algorithm 5. Note that the participating clouds do not communicate with each other and are possibly not aware concerning the number and identity of the other participants. Also note that all the clouds (including *reducer* and *randomizer*) need not to be reliable.

Initial Stage. The PC of each cloud is initially set by the dealer. The values of the PC are the secret shares of the first address of the client's program. In case there is no *randomizer* in the system, the dealer can guarantee that each cloud has enough precomputed values of polynomials to be used for degree reductions.

Execution Stage. In this stage, the clouds have to perform the following works:

- Program Execution: Each cloud executes the secret shared program independently and does not communicate with other clouds.
- Degree Reduction: Each cloud performs Algorithm 4 to reduce the polynomial degree of the shares which increased during the Subleq procedure.

Memory Refresh. Although we decreased the polynomial degree of the shared secret before write, the operation WRITE does increase the polynomial degree by ℓ each time. Thus, the read/write part of memory needs to be refreshed at intervals (e.g., every ten WRITE operations). Note that this part of memory can be relatively small compared with the whole memory, so it will not lead to too much bandwidth usage.

In Fig. 3, we give the outline of the program execution. The communication between the clouds and the dealer, and the communication between the clouds and the *reducer*(s) are all bidirectional. The dealer sends the secret shares of the client program and receives and reconstructs the program results executed by clouds. Moreover, we can use more than one *reducer* in

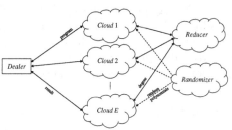

Fig. 3. The outline of Our RAM model

order to check the integrity of the results and identify which *reducer* is malicious.

Storage and Bandwidth. The storage of each cloud consists of the secret shares of the program instructions and the data. Notice that secret share of one bit needs one or multi-word size storage which is denoted by $\omega(1)$.

Data Table. Each row of the data table consists of the index and data item, it totally requires $(n + k)(n + k + t)\omega(1)$ words storage. As we previously assumed that the size of read-only table is much bigger than that of the read/write table, i.e., $k \gg n$, the storage requires roughly $O(k^2)\omega(1)$ words.

Instruction Table. The cloud stores an instruction table of size m, and each instruction consists of five parts. This requires $O(m)$ blocks storage with each block requires $O(3m + 2n + 2k)\omega(1)$ words.

Degree Reduction Table. According to the corresponding description of Algorithm 4, if a *randomizer* (or several *randomizers*) are used to produce secret shares of random polynomials on-line, no tables are needed. Otherwise every cloud needs to store a certain amount of shares which are pre-computed and dispatched by the dealer. These values could be generated and managed by a special database. The size of this database is dependent on the execution length of the program, i.e., about $O(mt\ell)\omega(1)$ words.

Bandwidth. For each Subleq, the clouds need to reduce the polynomial degrees of their data twice via communication with the *reducer* (and the *randomizer*). For each degree reduction from d to d^*, every cloud first obtains two shared evaluations from the *randomizer*, and then sends the *reducer* one word and receive $d-d^*$ coefficients from it, resulting in a total of approximately $O(k + m + t)\omega(1)$ words bandwidth used per cloud for one Subleq. In addition, the read/write memory needs to be refreshed at interval, it will result in $O(kt)\omega(1)$ words bandwidth usage. Therefore, in the worst case, the bandwidth of each cloud is $O(kt)\omega(1)$.

Security Analysis Sketch. We note that during the whole procedure of our model, all of the information are secret shared in E clouds and no original information will be leaked to the cloud itself. Besides this, our model has two characteristics:

Security Against Adversary Eavesdropping. For every LOAD operation, we had to compare the values stored in *PC* with all the indices in program table. It "touches" every position in the program table. Even through the adversary could eavesdrop on all the contents of *PC*, registers, etc., the adversary could not know which instruction in the table was executed. The same thing also happens in read/write operations. The characteristic is similar to the schemes that are based on fully homomorphic encryption, but here is information-theoretically secure.

Security Against Malicious Clouds. The malicious clouds include malicious participants and malicious *randomizer* and *reducer*. Informally, a malicious server can corrupt data in storage; and deviate from the prescribed protocol, particularly, not performing execution correctly.

For the participants: note that the program is outsourced to E clouds. Even if some of them output the wrong answers, the client can compare the results interpolated from different set of answers and obtain the correct result, or better off, use [30].

For the *reducer* and *randomizer*: every cloud may record the communication with the *randomizer* and *reducer* for audit, revealing possible malicious *reducers*. A possible strategy is to use several *reducers* simultaneously. After each cloud received the answers from the *reducers*, they could compare these results and notified the client/dealer whether the *reducers* were malicious or not. Similarly the actions of the *randomizer* can be monitored, say by forwarding the values sent by the *randomizer* to the *reducer*, requesting to the *reducers* to reveal all coefficients, and not use these values, requesting new values from the *randomizer*.

Unary vs. Binary. In our scheme, we use the unary representation for the instruction and data addresses. This type of representation is inappropriate if the clients program is very large because of its redundant bits. In a secret shared form, we have to use n words to represent these n bit which will lead to many operations over $\mathbb{F}_p$. As described in Sect. 2, we can use binary representation as a substitution. Compared with unary representation, binary representation can express exponentially more numbers with the same number of bits. However, using binary representation to perform secret string matching is more complicated and will require more degree reduction operations. In practical implementation, one can choose the representation based on the consideration of their memory and computation capacity.

5   Conclusions

We discussed a novel model for outsourcing arbitrary computations that provide confidentiality, integrity, and verifiability. Unlike the former RAM-based secure computation models, our scheme hides the client program and data all the time and manipulates the secrets directly. Therefore, no confidential information would be revealed. The setting is particularly interesting in the scope of big data that is stored in secret sharing fashion over the clouds, and there is a need to repeatedly compute functions over the data without reconstructing the data from the shares.

An important observation is that the dealer (and *reducer*(s)) may share common roots of all polynomials, unknown to the participating clouds, where addition and multiplications keep the roots unchanged. These unknown roots can serve as additional keys, the number of possible roots grows exponentially with the degree of the polynomials. Furthermore, implementation of interactive program is possible by reading and writing specific memory locations during the execution. Lastly, using several RISC instructions instead of OISC is possible to implement the program obliviously. For every instruction execution, we can perform each instruction once and using secret string match technique to ensure the right execution.

References

1. Amazon Elastic Compute Cloud (EC2). http://aws.amazon.com/ec2
2. Afshar, A., Hu, Z., Mohassel, P., Rosulek, M.: How to efficiently evaluate RAM programs with malicious security, Cryptology ePrint Archive, Report 2014/759 (2014)
3. Becker, G.T., Regazzoni, F., Paar, C., Burleson, W.P.: Stealthy dopant-level hardware trojans. In: Bertoni, G., Coron, J.-S. (eds.) CHES 2013. LNCS, vol. 8086, pp. 197–214. Springer, Heidelberg (2013)
4. Ben-Or, M., Goldwasser, S., Wigderson, A.: Completeness theorems for non-cryptographic fault-tolerant distributed computation. In: Proceedings of the Twentieth Annual ACM Symposium on Theory of Computing (STOC 1988), NY, USA, pp. 1–10. ACM, New York (1988)
5. Boyle, E., Goldwasser, S., Tessaro, S.: Communication locality in secure multi-party computation. In: Sahai, A. (ed.) TCC 2013. LNCS, vol. 7785, pp. 356–376. Springer, Heidelberg (2013)
6. Boyle, E., Chung, K.M., Pass, R.: Large-scale secure computation, Cryptology ePrint Archive, Report 2014/404 (2014)
7. Brenner, M., Wiebelitz, J., von Voigt, G., Smith, M.: Secret program execution in the cloud applying homomorphic encryption. In: Proceedings of the 5th IEEE International Conference on Digital Ecosystems and Technologies Conference (DEST), pp. 114–119 (2011)
8. Brenner, M., Perl, H., Smith, M.: How practical is homomorphically encrypted program execution? An implementation and performance evaluation. In: IEEE 11th International Conference on Trust, Security and Privacy in Computing and Communications (TrustCom), pp. 375–382 (2012)
9. Clash of the clouds. The Economist. http://www.economist.com/displaystory.cfm?story_id=14637206;2009
10. Damgård, I., Meldgaard, S., Nielsen, J.B.: Perfectly secure oblivious RAM without random oracles. In: Ishai, Y. (ed.) TCC 2011. LNCS, vol. 6597, pp. 144–163. Springer, Heidelberg (2011)
11. Dolev, S., Garay, J., Gilboa, N., Kolesnikov, V.: Swarming secrets. In: 47th Annual Allerton Conference, pp. 1438–1445 (2009)
12. Dolev, S., Gilboa, N., Li, X.: Accumulating automata and cascaded equations automata for communicationless information theoretically secure multi-party computation. In: Proceedings of the 3rd International Workshop on Security in Cloud Computing (SCC 2015), pp. 21–29. ACM, New York (2015)

13. Google Cloud Platform. https://cloud.google.com/storage/
14. Gentry, C.: Fully homomorphic encryption using ideal lattices. In: Proceedings of the 41st Annual ACM Symposium on Theory of Computing, pp. 169–178. ACM (2009)
15. Gentry, C.: A fully homomorphic encryption scheme, Ph.D. dissertation, Stanford University (2009)
16. Gentry, C., Halevi, S.: Implementing Gentry's fully-homomorphic encryption scheme. In: Paterson, K.G. (ed.) EUROCRYPT 2011. LNCS, vol. 6632, pp. 129–148. Springer, Heidelberg (2011)
17. Gentry, C., Goldman, K.A., Halevi, S., Julta, C., Raykova, M., Wichs, D.: Optimizing ORAM and using it efficiently for secure computation. In: De Cristofaro, E., Wright, M. (eds.) PETS 2013. LNCS, vol. 7981, pp. 1–18. Springer, Heidelberg (2013)
18. Goldreich, O.: Towards a theory of software protection and simulation by oblivious RAMs. In: STOC (1987)
19. Goldreich, O., Ostrovsky, R.: Software protection and simulation on oblivious RAMs. J. ACM **43**, 431–473 (1996)
20. Goodrich, M.T., Mitzenmacher, M., Ohrimenko, O., Tamassia, R.: Oblivious RAM simulation with efficient worst-case access overhead. In: ACM Cloud Computing Security Workshop (CCSW) (2011)
21. HOMOMORPHIC ENCRYPTION. http://sites.nyuad.nyu.edu/moma/projects.html
22. Liu, C., Huang, Y., Shi, E., Katz, J., Hicks, M.: Automating efficient RAM-model secure computation. In: Proceedings of the 2014 IEEE Symposium on Security and Privacy (SP 2014), pp. 623–638. IEEE Computer Society, Washington, D.C. (2014)
23. Lu, S., Ostrovsky, R.: How to garble RAM programs? In: Johansson, T., Nguyen, P.Q. (eds.) EUROCRYPT 2013. LNCS, vol. 7881, pp. 719–734. Springer, Heidelberg (2013)
24. Mazonka, O., Kolodin, A.: A simple multi-processor computer based on subleq, arXiv preprint arxiv:1106.2593 (2011). http://da.vidr.cc/projects/subleq/
25. Shamir, A.: How to share a secret. Commun. ACM **22**(11), 612–613 (1979)
26. Stefanov, E., Shi, E.: Multi-cloud oblivious storage. In: Proceedings of the 2013 ACM SIGSAC Conference on Computer and Communications Security (CCS 2013), NY, USA, pp. 247–258. ACM, New York (2013)
27. SUBLEQ. http://mazonka.com/subleq/
28. Tehranipoor, M., Koushanfar, F.: A survey of hardware trojan taxonomy and detection. IEEE Des. Test Comput. **27**(1), 10–25 (2010)
29. Wang, X., Huang, Y., Chan, T.-H.H., Shelat, A., Shi, E.: SCORAM: oblivious RAM for secure computation. In: The 21st ACM Conference on Computer and Communications Security (CCS), Scottsdale, Arizona, USA, November 2014
30. Welch, L., Berlekamp, E.R.: Error correction for algebraic block codes, US Patent, 4 633 470 (1983)
31. Zhuravlev, D., Samoilovych, I., Orlovskyi, R., Bondarenko, I., Lavrenyuk, Y.: Encrypted program execution. In: IEEE 13th International Conference on Trust, Security and Privacy in Computing and Communications (TrustCom), pp. 817—822 (2014)

Column Generation Integer Programming for Allocating Jobs with Periodic Demand Variations

Ikbel Belaid[1,2](✉) and Lionel Eyraud-Dubois[1,2]

[1] Inria Bordeaux – Sud-Ouest, Talence, France
[2] University of Bordeaux, Talence, France
{Ikbel.Belaid,Lionel.Eyraud-Dubois}@inria.fr

Abstract. In the context of service hosting in large-scale datacenters, we consider the problem faced by a provider for allocating services to machines. An analysis of a public Google trace corresponding to the use of a production cluster over a long period shows that long-running services experience demand variations with a periodic (daily) pattern, and that services with such a pattern account for most of the overall CPU demand. This leads to an allocation problem where the classical Bin-Packing issue is augmented with the possibility to co-locate jobs whose peaks occur at different times of the day, which is bound to be more efficient than the usual approach that consist in over-provisioning for the maximum demand. In this paper, we propose a column-generation approach to solving this problem, where the subproblem uses a sophisticated SOCP (Second Order Cone Program) formulation. This allows to explicitly select jobs which benefit from being co-allocated together. Experimental results comparing with theoretical lower bounds and with standard packing heuristics shows that this approach is able to provide very efficient assignments in reasonable time.

1   Introduction

The Cloud paradigm provides an illusion of infinite elasticity and seamless provisioning of IT resources. However, as providers keep scaling their infrastructures year after year, the efficient allocation of services in *Platform-as-a-Service* (PaaS) becomes crucial.

We concentrate on the case of a Cloud platform in which several independent services, typically virtualized as Virtual Machines (VMs) or lightweight containers, are serving user queries and need to be allocated onto physical machines (PMs) [1,19]. We consider the static case where a set of *dominant* services define the overall resource usage of the physical platform, which has proved to be commonplace in large datacenters [3]. In this context, mapping services with heterogeneous computing demands onto PMs is amenable to a multi-dimensional Bin-Packing problem (each dimension corresponding to a different kind of resource, memory, CPU, disk, bandwidth,...). Indeed, on the infrastructure side, each physical machine presents a given computing capacity (*i.e.* the number of Flops

© Springer International Publishing Switzerland 2016
I. Karydis et al. (Eds.): ALGOCLOUD 2015, LNCS 9511, pp. 35–48, 2016.
DOI: 10.1007/978-3-319-29919-8_3

it can process during one time-unit), a memory capacity and a failure rate
(*i.e.* the probability that the machine will fail during the next time period).
On the client side, each service has a set of requirements along the same dimensions (memory and CPU footprints) and a reliability requirement that has been
negotiated typically through an SLA [9].

In this work, we consider a specific feature of CPU demand that arises in the
context of service allocation. Previous work on the subject [4] argues that many
services representing most of the overall CPU demand exhibit daily patterns and
their demand can be modeled as a set of sinusoids, each comprising a constant
component, an amplitude and a phase. This premise gives rise to a model for
jobs with time-varying resource demands and to the associated packing problem.
Such a model can be used to aggregate onto the same physical machines more
resources than it would be possible based on their maximal demands only, taking
advantage of the fact that different phases for different services imply that peak
demands do not occur simultaneously. In this paper, we propose an algorithm
based on column generation for packing jobs with periodic demands on the
hosting platform. This algorithm provides very efficient allocations, compared
to state-of-the-art greedy packing heuristics.

The remaining of this paper is organized as follows. We discuss some related
works in Sect. 2. In Sect. 3, we present the formulation of the optimization problem as a *Second Order Cone Program* (SOCP). In Sect. 4, we propose our efficient
packing algorithm based on column generation, whose performance is analyzed
and validated on realistic and simulated data in Sect. 5. Finally, conclusions are
drawn in Sect. 6.

2   Related Works

In order to deal with resource allocation problems arising in the context of
Clouds, several sophisticated techniques have been developed in order to optimally allocate user services onto PMs, either to achieve good load-balancing [5, 8]
or to minimize energy consumption [6]. Most of the approaches in this domain
are based on offline [10] and online [11] variants of Bin-Packing strategies.

In this paper, we concentrate on the allocation of jobs that last for a long
time and whose CPU demands exhibit periodic patterns. Some other work deal
with allocating jobs whose demands varies over time, either with predictable
(static) or unknown (dynamic) behavior. In the static case which is the focus
of this present work, historical average resource utilization is typically used as
input to an algorithm that maps services to physical machines. Therefore, the
mapping is done off-line. In contrast, dynamic allocation schemes are implemented on shorter timescales. Dynamic allocation leverages the ability to perform runtime migrations of jobs and to recompute resource allocation amongst
services. A dynamic migration algorithm *Measure Forecast Remap* is introduced
in [7], where highly variable workloads are forecast over intervals shorter than
the time scale of demand variability to ensure dynamic execution minimization
of the number of required machines. Based on stochastic vector packing model,

the static scheme proposed in [15] makes use of customers' periodic access patterns in web server farms to assign each customer to a server so as to minimize the total number of required servers. In this latter work, the variable demand is analyzed at a different time scale to extract probability distributions that are independent of time. Then, *stream-packing* heuristics are employed to select the most complementary jobs to be packed in the same server. Urgaonkar et al. [16] rely on on-line application profiling to demonstrate the feasibility and benefits of overbooking resources in shared platforms to guide the application placement onto dedicated resources while providing performance guarantees at runtime. A new mechanism for dynamic resource management in cluster-based network servers [2], called cluster reserve, allows performance isolation between service classes and provides a minimal amount of resources, irrespective of the load imposed by other requests. In contrast to these other directions, our work focuses on a part of the workload which exhibits deterministic periodic variability. In this context, dynamic resource management is unnecessary: the migration cost can be avoided by using periodicity-aware static approaches for service allocation. By focusing on long-running services with high workloads, it is possible to apply sophisticated techniques to provide efficient packing policies, which results in increased resource usage. Smaller or short-lived jobs, which are much more numerous but represent a smaller part of the resource usage, can be handled with usual greedy allocation schemes.

This paper is a followup to [4], which analyzes the performance of several standard packing heuristics in the context of packing jobs with periodic demand variation. In this paper, we propose the use of the *Dantzig-Wolfe* decomposition [17] to solve very efficiently the corresponding packing problem. In fact, mathematical programs featured by a large space of integer variables are particularly suited for *Dantzig-Wolfe* decomposition that reformulates the original compact problem to provide a tighter linear programming relaxation bound. This decomposition relies on delayed column generation algorithm. The overarching idea of this algorithm is that many programs are too large to consider all the variables explicitly. Since most of the variables will be neglected in the optimal solution, only a subset of variables need to be considered in theory when solving the problem. Column generation leverages this idea to generate only the variables which have the potential to improve the objective function, that is, to find variables with negative reduced costs. Section 4 details the utilisation of *Dantzig-Wolfe* decomposition to reformulate the packing of jobs with variable demands on hosted parallel machines based on the original formulation and employing the column generation algorithm.

The *Dantzig-Wolfe* reformulation gives rise to a master problem and subproblems, whose typically large number of variables is dealt with implicitly by using an integer programming column generation procedure, also known as branch-and-price algorithm. Solving the master problem does not require an explicit enumeration of all its columns because the column generation algorithm allows one to generate columns if and when needed. In many cases, this allows huge integer programs that had been previously considered intractable to be

solved. The technique of *Dantzig-Wolfe* using the approach of column generation has been applied successfully in many classical problems as: cutting stock,
vehicle routing, crew scheduling, etc.

3   Packing of Jobs with Periodic Demands

An analysis of a publicly available Google trace [13,14] has shown that about
two-thirds of the dominant, normal production jobs in that trace exhibit significant daily pattern [3], and that their demand can be approximated with
sinusoidal functions. Based on this analysis, and following [4], we consider a
packing problem for those long running jobs, which account for a large portion
of the workload.

3.1   Notations and Problem Formulation

Let us assume that the cloud platform we consider consists of M homogeneous
nodes $M_1, \ldots, M_k, \ldots, M_M$ and let us denote the processing capacity of a node
by C. For the sake of simplicity and in order to focus on issues related to the
aggregation of periodic demands, we will concentrate on CPU demands only.
The tasks of a job (corresponding to a service in the trace) can run on any node,
and job J_j is split into N_j tasks denoted by $T_{j,1}, \ldots, T_{j,l}, \ldots, T_{j,N_j}$, who share
the same characteristics in terms of CPU demand.

   In turn, platform nodes are allowed to run several tasks, provided that at any
time, their capacity is not exceeded. We assume that the set of tasks running on
a node does not change over time, what is a realistic assumption for dominant
Normal Production jobs, and we model the instantaneous demand at time t of
task $T_{j,l}$, which does not depend on l, as

$$W_j(t) = C_j + \rho_j \sin\left(2\pi \frac{t}{P} + \phi_j\right)$$

where C_j denotes the average of CPU demand of Task $T_{j,l}$, ρ_j denotes the
maximal amplitude of the variation of the demand, ϕ_j denotes its phase, and P
denotes the common period for all jobs.

   In this context, our aim is to provide a static packing for the set of tasks $T_{j,l}$
such that at any step and on any resource, capacity constraints are not exceeded
and such that the number of required nodes is minimized. More specifically, this
model allows to take advantage of daily variations in order to obtain an efficient
packing of tasks. Indeed, most packing strategies are based on the maximal
demand of each task, what corresponds to $C_j + \rho_j$ for a task of job j. Taking
advantage of the fact that all tasks do not achieve their peak demand at the
same time in the day, it is possible to pack more tasks, and therefore to use
fewer nodes whilst packing statically all the tasks.

   The corresponding capacity constraint for a given machine M_k is thus

$$\forall t, \quad \sum_{j,l:\ T_{j,l} \in M_k} W_j(t) \leq C,$$

and it can be rewritten [4] as an expression which does not depend on t:

$$\forall k, \quad \sum_{j,l:\ T_{j,l}\in M_k} C_j + \sqrt{(\sum_{j,l:\ T_{j,l}\in M_k} \rho_j \cos(\phi_j))^2 + (\sum_{j,l:\ T_{j,l}\in M_k} \rho_j \sin(\phi_j))^2} \leq C$$

(1)

This modified packing constraint yields a quadratically constrained programming (QCP) formulation of the problem. This formulation uses two types of variables: integer variables $X_{j,k}$ representing the number of tasks of job j allocated on the node M_k, and boolean variables Y_k representing whether node N_k is used. With these variables, the formulation is the following:

$$\text{Minimize} \sum_k Y_k$$

$$\forall j \in J, \quad \sum_{k\in M} X_{j,k} = N_j$$

(2)

$$\forall k \in M, \quad (\sum_{j\in J} X_{j,k}\ \rho_j\ \cos(\phi_j))^2 + (\sum_{j\in J} X_{j,k}\ \rho_j\ \sin(\phi_j))^2$$

$$\leq (C\ Y_k - \sum_{j\in J} X_{j,k}\ C_j)^2$$

(3)

$$\forall k \in M, \quad C\ Y_k - \sum_{j\in J} X_{j,k}\ C_j \geq 0$$

(4)

In this formulation, constraint (2) ensures that all instances of all jobs are allocated. Tasks belonging to the same job could co-exist in the same node. Constraints (3) and (4) are a quadratic reformulation of Eq. (1), ensuring that an unused node does not contribute any resource to the platform. Due to the nature of this constraint, this formulation can be expressed as a Second Order Cone Program (SOCP) [12], and can thus benefit from efficient general purpose solvers [12] for convex optimization. However, as noticed in [4], on real-size instances with thousands of machines, this formulation can not be solved in reasonable time with integer and boolean values. Relaxing the problem by allowing rational variables makes it possible to obtain a lower bound on the necessary number of resources in reasonable time.

In the next Section, we describe how to reformulate this problem with a *Dantzig-Wolfe* decomposition, which allows to quickly obtain very good solutions to the packing problem.

4 Integer Programming Column Generation

Dantzig-Wolfe decomposition has been an important tool to solve large structured models that could not be solved using standard algorithms as they exceeded the capacity of solvers. The main idea behind this technique is to decompose the original problem into a number of independent subproblems, whose solutions are then assembled by solving a so-called master problem. This master problem is then solved iteratively. In our case, we can identify the natural decomposition of the problem: for two different values of k (i.e. for each machine), the corresponding sets of constraints (3) and (4) are independent, because they contain disjoint sets of variables. Since we assume that machines are homogeneous, all those subproblems are actually identical, and we obtain a special case where solving it only once is enough.

In the *Dantzig-Wolfe* reformulation, we obtain a master problem which contains one variable for each solution to this subproblem. In our case, such a solution is simply a valid packing configuration for a machine, i.e. a set of jobs which can be allocated together on a single machine while respecting the capacity constraint. One configuration Z_i can be represented as a J-uplet $(X_{1,i}, X_{2,i}, \ldots, X_{J,i})$, where $X_{j,i}$ is the number of tasks of job j in configuration i. As discussed previously, this configuration is valid if it satisfies Eq. (1).

In the following, we will denote as K the set of all valid configurations. The master problem contains one variable Y_i for each configuration $Z_i \in K$, which represents the number of machines which use the configuration Z_i, i.e. the number of machines to which $X_{j,i}$ tasks of job j are allocated. The packing problem can now be formulated as follows:

Master Problem: Minimize $\sum_{i \in K} Y_i$ s.t

$$\forall j \in J, \quad \sum_{i \in K} X_{j,i} \, Y_i \geq N_j \tag{5}$$

$$Y_i \in \mathbb{N} \tag{6}$$

This master problem cannot be solved directly due to an exponential number of variables. However, the column generation approach consists in considering variables only from a subset $K' \subset K$, and to solve the resulting *restricted master problem* (RMP) on this set of variables.

This restricted problem may provide a sub-optimal solution, since there might exist a configuration in $K \setminus K'$ which improves the solution. In order to find this configuration, one can write the dual of the master problem, in which there is one variable π_j for each job, and one constraint for each configuration:

Dual Master Problem: Maximize $\sum_{j \in J} N_j \pi_j$ s.t

$$\forall i \in K, \quad \sum_{j \in J} X_{j,i} \, \pi_j \leq 1 \tag{7}$$

$$\pi_j \geq 0 \tag{8}$$

The sub-optimal solution obtained from the restricted master problem provides a solution π_j^* of the dual master problem which is possibly infeasible, since not all constraints are included in the dual of the RMP. From this solution π_j^*, we can identify a variable to add to the problem by searching for a violated constraint in the dual, i.e. a configuration $Z_i \in K$ such that $\sum_{j \in J} X_{j,i} \, \pi_j^* > 1$. This gives rise to the following subproblem, with one variable U_j for each job:

Subproblem: Periodic Knapsack: Minimize $1 - \sum_{j \in J} \pi_j^* U_j$ s.t

$$\forall j \in J, \quad \left(\sum_{j \in J} U_j \, \rho_j \, \cos(\phi_j)\right)^2 + \left(\sum_{j \in J} U_j \, \rho_j \, \sin(\phi_j)\right)^2 \leq \left(C - \sum_{j \in J} U_j \, C_j\right)^2 \quad (9)$$

$$U_j \in \mathbb{N} \tag{10}$$

If the optimal solution of this subproblem has a negative value, then we have identified a configuration to add to the RMP. On the other hand, if this problem has no solution with negative value, it means that all constraints in the dual of the master problem are satisfied with the current solution, which implies that this current solution of the RMP is actually optimal for the master problem.

This subproblem can be seen as a knapsack problem: given profits π_j^* for each job, we search for the set of tasks with maximal profit which can fit in a single machine. The strong point of the *Dantzig-Wolfe* reformulation in this case is that we have isolated the quadratic constraint in the subproblem, which is of much smaller scale than the original problem, with only one variable per job. This problem can now be solved (quite efficiently as we will see in Sect. 5) with a general integer SOCP solver.

The column generation algorithm is summarized in Algorithm 1: starting from an initial set of configurations (which we describe below), the algorithm iteratively solves the RMP, and then uses the values of the dual variables as prices in the knapsack subproblem. The solution to this subproblem yields a new configuration which is added to the set, and a new iteration is performed. The process ends when no solution to the subproblem has a negative cost. The obtained RMP is then solved with integer constraints to obtain a feasible solution to the original problem. In practice, this last step is often too time consuming for the solver to obtain an optimal solution in reasonable time, so in our experimental evaluation, we included a 5 min time limit and use the best feasible solution obtained by the solver in that time.

The initial set of configurations can be chosen arbitrarily, as long as the first RMP is feasible, i.e. all jobs are represented in at least one configuration. For simplicity, we build the initial set K_0 with one configuration per job, where the configuration for job j contains $\left\lfloor \frac{C}{C_j + \rho_j} \right\rfloor$ tasks of job j (as many as can fit on one machine), and 0 tasks of all other jobs.

5   Experimental Evaluation

In this section, we present the results of synthetic and realistic experiments provided by column-generation algorithm and best-effort heuristics. We investigate the performance of each in reducing the number of used nodes as well as their margin towards the lower bound. We show that our column generation algorithm delivers good results in reducing the number of iterations and computation time.

Data: Job characteristics: C_j, ρ_j, ϕ_j and N_j
Result: Feasible solution: a set K_t of configurations and values $(Y_i)_{i \in K_t}$ stating
         how many machines use each configuration
$t \leftarrow 0$
$K_t \leftarrow K_0$
repeat
  | Solve RMP with variables in K_t
  | Generate dual values π_j^* from the RMP solution
  | Solve the subproblem with prices π_j^*
  | **if** *strictly negative reduced cost* **then**
  | | $Col \leftarrow$ new column with the coefficients of the subproblem solution
  | | $t \leftarrow t + 1$
  | | $K_t \leftarrow K_{t-1} \cup Col$
  | **end**
until *no negative reduced cost solution*;
Solve RMP with variables in K_t as an integer program

Algorithm 1. Column generation algorithm

5.1 Complexity and Lower Bound

The optimization problem that consists in packing tasks with periodic demands into nodes is clearly NP-Complete, since it is amenable to classical Bin-Packing problems [10,11] in its most simplified setting where $\forall j$, $\rho_j = 0$, *i.e.* the case when demands do not change over time. Indeed, in many cases the last step of our column generation algorithm (where we look for an integer solution) is unable to obtain an optimal solution in the alloted time. In order to asses the performance of the obtained results, we rely on a simple but powerful lower bound: the total workload at time t is $\sum_{j \in J} W_j(t)$, whose peak can be computed like in Sect. 3 as $W = \sum_{j \in J} C_j + \sqrt{(\sum_{j \in J} \rho_j \cos(\phi_j))^2 + (\sum_{j \in J} \rho_j \sin(\phi_j))^2}$. Since in any solution, the sum of the capacity used on each machine is not lower than P, we know that any solution must use at least $\lceil \frac{W}{C} \rceil$ machines. This solution is not feasible in general but it provides a lower bound on the number of necessary nodes.

5.2 Heuristics

In this Section, we present some heuristics introduced in [4], adapted from classical efficient greedy Bin-Packing algorithms to the case of tasks exhibiting daily patterns. In the following, we denote by $\mathcal{L}(M_k, T_{j,l})$ the peak load on machine M_k after adding one task $T_{j,l}$ of job J_j to M_k.

- Best-Fit Decreasing $\mathcal{BFD}$ is a greedy algorithm in which tasks are considered by decreasing values of C_j. At any step, task $T_{j,l}$ (from job J_j) is allocated to the node M_k such that $\mathcal{L}(M_k, T_{j,l})$ is maximized (while remaining below C). Note that contrarily to what happens in classical $\mathcal{BFD}$, the size that is considered is the size after the allocation. If no such node exists, then a new node is added to the system to hold the task.

- In Min-Max $\mathcal{MM}(M)$, the target number of nodes is fixed to M a priori. Then, $\mathcal{MM}$ is a greedy algorithm where tasks are considered by decreasing values of C_j. At any step, task $T_{j,l}$ (from job J_j) is allocated to the node M_k such that $\mathcal{L}(M_k, T_{j,l})$ is minimized, in order to balance the load between the different nodes. The allocation may fail if M is too small. We thus use dichotomic search to find the smallest value of M which allows to obtain a solution.
- Min-Max-Module $\mathcal{MMM}$ is similar to $\mathcal{MM}$, except that tasks are represented using their maximal demand over time $C_j + \rho_j$ only. This is typically what happens when one neglects the possibility to take advantage of the fact that peak demands do not occur at the same time for all jobs.

5.3   Simulated Synthetic Data

First, we perform a set of experiments with synthetic data in order to assess the influence of the parameters on the performance of the different proposed methods. In all the experiments, we set the capacity of the nodes to 20, and we display the ratio between the number of nodes provided by the corresponding method against the lower bound on the number of necessary nodes described in Sect. 5.1.

For synthetic jobs, we consider the following parameters:

- CPU footprint of the tasks: we consider the case of Large Tasks, Medium Tasks and Small Tasks where C_j is chosen uniformly at random respectively in $[0, 10]$, $[0, 5]$ and $[0, 1]$. To keep the total workload constant, we set the number of tasks N_j in each job to 50 for Large Tasks, 100 for Medium Tasks, and 500 for Small Tasks.
- Daytime amplitude: we consider the case of Large Daytime Amplitude (where ρ_j is chosen uniformly at random in $[0, C_j]$) and Small Daytime Amplitude (where ρ_j is chosen uniformly at random in $[0, \frac{C_j}{2}]$).

In all cases, the phase of each job is chosen uniformly at random in $[0, 2\pi]$, and the number of jobs is set to 100. We performed other experiments with different number of jobs and tasks, but the results showed very little sensitivity to these parameters and were excluded from the paper in order to save space. For each scenario, we have performed 20 experiments, and the results are shown on Fig. 1, where the 20 experiments for each scenario and each algorithm are grouped together in a boxplot showing the mean, the first and third quantiles, and minimum and maximum values.

The figure shows that the column generation algorithm is able to consistently provide solutions with a number of required nodes very close to the lower bound, in all of the scenarios. Actually, for some of the Large Tasks scenarios, the solver is able to obtain a provably optimal integer solution in the final step of the algorithm, meaning that the column generation algorithm actually provides an optimal solution in these cases. This also shows that the lower bound is actually very precise, since it is possible to exhibit a feasible solution with very close performance.

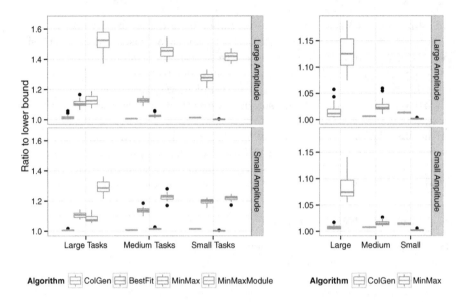

Fig. 1. Performance of all algorithms on synthetic data. The right plot is a focus on the most efficient algorithms.

In the Small Tasks scenarios, in which each node can hold a few tens of tasks, Min-Max $\mathcal{MM}$ performs extremely well and is always at most within 1 % of the lower bound. This behaviour is usual in Bin-Packing problems: the presence of very small objects makes the packing easier since they can be used to fill the wasted space in the bins. Indeed, the results of Min-Max $\mathcal{MM}$ degrade when tasks get larger: in this case, the number of tasks per node is relatively small (a few units) and greedy heuristics fail to achieve close to optimal performance. On the other hand, in the Small Tasks scenarios, the solution provided by the column generation algorithm is not as good. This comes from the fact that the integer problem of the final step is very difficult to solve in that case.

Nevertheless, the number of nodes required by the column generation algorithm is always within 1 % of the lower bound. It is worth noting that in the context which we consider in this work (long-running services with heavy workload), the task sizes are not small, and the medium-large task sizes are the most realistic cases (see Sect. 5.4 for a comparison on a real trace).

The BestFit heuristic $\mathcal{BFD}$ represents the standard packing algorithm used in such Cloud systems. We can see that its performance is consistently 10 % above the lower bound, and even worse in the case of small tasks. This advocates strongly in favor of more sophisticated algorithms like the one we propose. Finally, we can also see that failing to take periodic demand variations leads to a large waste of resources. Indeed, the performance of Min-Max-Module $\mathcal{MMM}$ is consistently far from the lower bound, by 50 % in the case of Big Amplitudes and by 25 % in the case of Small Amplitudes.

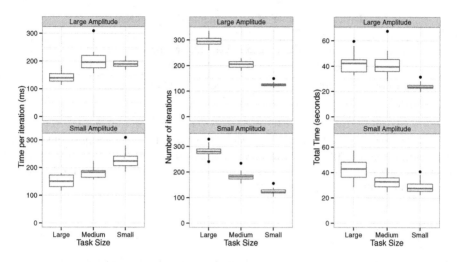

Fig. 2. Running time of column generation algorithm.

Figure 2 analyses the computation time for the first phase of the column-generation algorithm, which solves the rational relaxation of the problem (the second phase is the final step where we obtain an integer feasible solution, whose time is limited to 5 min in our experiments). We observe that the number of iterations remains below 300, and the time per iteration is very low (around 200 ms), showing that the SOCP formulation for the subproblem is very efficient. This allows the column generation algorithm to complete its first phase in about 40 s in all scenarios.

5.4   Jobs and Tasks of Google Trace

Then, we concentrate on the set of realistic periodically variable jobs in the trace released by Google [18] and corresponding to one production center. In [4], a instance has been extracted from this trace with 89 jobs corresponding to a total of 22600 tasks. The largest job in terms of tasks consists of 1608 tasks and the largest job in terms of CPU demand corresponds to the capacity of 184 nodes at its peak demand. A capacity equivalent to 2198 nodes would be required if all jobs reached their peak demand at the same instant. The overall peak demand for the whole set of jobs is equivalent to the capacity of 2090 nodes. Therefore, there exists a potential improvement on the number of required nodes of 5 %, what should be considered as large in the context of an actual production center. We have applied our column-generation algorithm on this instance, and the results achieved are displayed in Table 1.

The results of $\mathcal{MM}$ are deemed extremely good in [4], because the number of required machines is only 1.1 % higher than the lower bound. Our column-generation algorithm $\mathcal{CG}$ is able to provide an even more efficient solution, with a number of nodes only 0.6 % higher than the lower bound, effectively halving

Table 1. Number of nodes required per heuristic.

	Column generation $\mathcal{CG}$	Best-Fit $\mathcal{BFD}$	Min-Max $\mathcal{MM}$	Min-Max-Module $\mathcal{MMM}$
Number of nodes	2103	2182	2114	2226

the gap between the best solution and the lower bound. As shown previously, the time complexity of our algorithm is very reasonable, showing that our column generation algorithm can really make an impact for improving resource usage in actual production centers.

6 Conclusions

Allocating computing resources for multiple time-varying job workloads is an attractive yet challenging target for many providers of large-scale infrastructures of cloud computing. Towards this end, we address in this paper a resource allocation problem for jobs that exhibit daily periodic sinusoidal patterns. Such jobs have been shown to represent a significant part of the workload of large production clusters, as exemplified by a trace from a Google center. Taking the periodic pattern into account allows to coallocate jobs which reach their peaks at different times, and this allows to significantly increase the resource usage on these platforms.

In this paper, we present a novel packing technique relying on job aggregation mechanism by employing an exact method using column generation integer programming. The *Dantzig-Wolfe* reformulation allows to isolate the quadratic constraint in a small size subproblem, which can be solved very efficiently. Solving iteratively this subproblem and the reformulated packing problem allows to efficiently identify the relevant machine configurations, i.e. the set of jobs which should be allocated together. This technique is then compared to best-effort heuristics inspired from the standard bin-packing methods, on both simulated and realistic data. Experimental results show that this algorithm obtains good results very consistently, even in difficult cases in which task sizes are large and few tasks can fit together on the same machine. In the most realistic cases, the column generation improves over the best heuristic by up to 5 %, effectively halving the gap between the best known solutions and the lower bound.

As future work, we plan to extend job aggregation strategies to provide performance guarantees for other resources like memory, disk, network bandwidth, etc. Improving the column generation algorithm could focus on two different directions: using a more efficient routine to solve the subproblem could lower further the running time of the first phase, and more efficient branching schemes could improve the efficiency of the last step of the algorithm. Besides, we target to address the problem of resource allocation and sharing for dynamically arriving jobs while considering the already assigned static ones. This problem is challenging and attractive computing paradigm in cloud computing for a wide variety of applications. This dynamic co-allocation for unpredictable jobs presents new challenges to resource management in multicluster systems, such as locating sufficient resources for these dynamic jobs in distributed sites, managing temporarily the job assignment and coordinating their executions with the processing of the static jobs.

References

1. Armbrust, M., Fox, A., Griffith, R., Joseph, A.D., Katz, R.H., Konwinski, A., Lee, G., Patterson, D.A., Rabkin, A., Stoica, I., et al.: Above the clouds: a Berkeley view of cloud computing, University of California, Berkeley (2009)
2. Aron, M., Druschel, P., Zwaenepoel, W.: Cluster reserves: a mechanism for resource management in cluster-based network servers. In: Proceedings of the ACM SIG-METRICS Conference, pp. 90–101 (2000)
3. Beaumont, O., Eyraud-Dubois, L., Lorenzo-Del-Castillo, J.-A.: Analyzing real cluster data for formulating allocation algorithms in cloud platforms. In: 2014 IEEE 26th International Symposium on Computer Architecture and High Performance Computing (SBAC-PAD), pp. 302–309 (2014)
4. Beaumont, O., Belaid, I., Eyraud-Dubois, L., Lorenzo-Del-Castillo, J.-A.: Allocating jobs with periodic demand variations. EuroPar **2015**, (February 2015)
5. Beaumont, O., Eyraud-Dubois, L., Rejeb, H., Thraves, C.: Heterogeneous resource allocation under degree constraints. IEEE Trans. Parallel Distrib. Syst. (2012)
6. Beloglazov, A., Buyya, R.: Energy efficient allocation of virtual machines in cloud data centers. In: IEEE/ACM International Conference on Cluster, Cloud and Grid Computing, pp. 577–578. IEEE (2010)
7. Bobroff, N., Kochut, A., Beaty, K.: Dynamic placement of virtual machines for managing SLA violations. In: 10th IFIP/IEEE International Symposium on Integrated Network Management, IM 2007, pp. 119–128 (2007)
8. Calheiros, R.N., Buyya, R., De Rose, C.A.F.: A heuristic for mapping virtual machines and links in emulation testbeds. In: Proceedings of International Conference on Parallel Processing (ICPP), pp. 518–525. IEEE (2009)
9. Cirne, W., Frachtenberg, E.: Web-scale job scheduling. In: Cirne, W., Desai, N., Frachtenberg, E., Schwiegelshohn, U. (eds.) JSSPP 2012. LNCS, vol. 7698, pp. 1–15. Springer, Heidelberg (2013)
10. Garey, M.R., Johnson, D.S.: Computers and Intractability, A Guide to the Theory of NP-Completeness. W. H. Freeman and Company, New York (1979)
11. Hochbaum, D.: Approximation Algorithms for NP-Hard Problems. PWS Publishing Company, Boston (1997)
12. Mittelmann, H.D.: An independent benchmarking of SDP and SOCP solvers. Math. Program. **95**(2), 407–430 (2003)
13. Reiss, C., Tumanov, A., Ganger, G.R., Katz, R.H., Kozuch, M.A.: Towards understanding heterogeneous clouds at scale: Google trace analysis. Technical report ISTC–CC–TR–12–101, Intel science and technology center for cloud computing, Carnegie Mellon University, Pittsburgh, PA, USA, April 2012. http://www.istc-cc.cmu.edu/publications/papers/2012/ISTC-CC-TR-12-101.pdf
14. Reiss, C., Wilkes, J., Hellerstein, J.L.: Google cluster-usage traces: format + schema. Technical report, Google Inc., Mountain View, CA, USA, November 2011. Revised 20 March 2012. http://code.google.com/p/googleclusterdata/wiki/TraceVersion2

15. Shahabuddin, J., Chrungoo, A., Gupta, V., Juneja, S., Kapoor, S., Kumar, A.: Stream-packing: resource allocation in web server farms with a QoS guarantee. In: Monien, B., Prasanna, V.K., Vajapeyam, S. (eds.) HiPC 2001. LNCS, vol. 2228, pp. 182–191. Springer, Heidelberg (2001)
16. Urgaonkar, B., Shenoy, P., Roscoe, T.: Resource overbooking, application profiling in shared hosting platforms. SIGOPS Oper. Syst. Rev. **36**(SI), 239–254 (2002)
17. Vanderbeck, F.: On dantzig-wolfe decomposition in integer programming and ways to perform branching in a branch-and-price algorithm. Oper. Res. 111–128 (2000)
18. Wilkes, J.: More Google cluster data. Google research blog, November 2011. http:// googleresearch.blogspot.com/2011/11/more-google-cluster-data.html
19. Zhang, Q., Cheng, L., Boutaba, R.: Cloud computing: state-of-the-art and research challenges. J. Internet Serv. Appl. **1**(1), 7–18 (2010)

SSSDB: Database with Private Information Search

Hillel Avni[1], Shlomi Dolev[1], Niv Gilboa[2], and Ximing Li[3(✉)]

[1] Department of Computer Science,
Ben Gurion University of Negev, Beer-Sheva, Israel
hillel.avni@gmail.com, dolev@cs.bgu.ac.il
[2] Department of Communication Systems Engineering,
Ben Gurion University of the Negev, Beer-Sheva, Israel
niv.gilboa@gmail.com
[3] College of Mathematics and Informatics,
South China Agricultutral Unversity, Guangzhou, China
liximing@scau.edu.cn

Abstract. This paper presents searchable secret shares (SSS), a novel method to search and collect statistics about private information quickly without retrieving secretly shared data, which is stored in public clouds separately. The new capabilities of SSS serve as a base for a newly defined SSS database SSSDB with reduced communication overhead and better security compared with private information retrieval (PIR) based databases. Clouds know neither the database stored in their disks nor the searched patterns and the results. Unlike PIR based databases, SSSDB does not need to maintain an access data structure for mapping keys to indexes performing inserts and deletes in the user memory, and therefore avoids possible information leakage. SSSDB is information-theoretically secure and can be implemented using a small integer field which implies high memory and computation efficiencies.

1 Introduction

Next generation IT infrastructures are based on private, public and hybrid cloud infrastructures. When private files (documents, pictures, mails) are stored in public clouds, cloud providers have full access to the data and control where it is stored. While they do not have much information about the infrastructure and the security mechanisms in place, in particular, the storage may be out of country, which could imply legal concerns. The lack of transparency concerning the storage settings in the cloud prevents security officers, working for big enterprise organizations, and managing sensitive data, from using cloud resources. As an example, we can take a health care provider that is constantly looking for new ways to reduce the cost of data centers and management operations is moving

X. Li—Supported by the National Natural Science Foundation of China (No. 61472146, 61402184), Science and Technology Planning Project of Guangdong Province (2013B010401020), the Project-sponsored by SRF for ROCS, SEM.

I. Karydis et al. (Eds.): ALGOCLOUD 2015, LNCS 9511, pp. 49–61, 2016.
DOI: 10.1007/978-3-319-29919-8_4

his/hers data centers to the cloud. Such a migration implies great savings, reliability and efficiency. Still, organizations have doubts since patients' data privacy is not guaranteed. In addition, once the database is migrated to the cloud, clients searching in the database also want to protect themselves from possible tracing, by cloud providers, of their (or their medical doctor) inquires concerning their health record. Many methods have been proposed to protect the privacy of the data provider and the data user.

1.1 Private Data Search in Clouds

Private information retrieval [3] (PIR) allows a client to look up information in an online database without letting the database servers learn the query terms or responses. An inherent limitation of PIR is that searching in a record requires retrieving the record. Retrieving secret shares for a record may incur significant communication. One way to avoid retrieval is to use fully homomorphic encryption (FHE). In his seminal paper [5], Craig Gentry presented the first FHE scheme which is capable of performing encrypted computation on Boolean circuits. A user specifies encrypted inputs to the program, and the server computes on the encrypted inputs without gaining information concerning the input or the computation state. Following the outline of Gentry's, many subsequent FHE schemes [6,7] are proposed, some of which are even implemented [8]. Recently, Craig Gentry et al. executed one AES-128 encryption homomorphically in eight days. Such overhead is unacceptable for most applications.

CryptDB [11] is a secure database implementation that uses several techniques including complex mechanisms such as order preserving encryption (OPE), which is computationally secure. The combination of the variety of mechanisms implies complicated security proof, attempting to identify the weak chain in the composition of computational security based techniques and verify the overall security. Secure searchable secret sharing schemes [13] base their security on oblivious RAM, which has poor performance and does not completely hide the searched item.

Some recent works [2,9,10,14] proposed private data searching schemes based on Shamir's secret sharing algorithm instead of single copy encryption. These schemes are mainly based on sorting shares by secret value, which reveals some information about the data.

In this paper we propose a method, which is based on pure secret sharing scheme, to search in the secret shares without retrieving them, with a complexity level similar to the best PIR based scheme. Security First Corp and Navajo Systems [1] use complicated secret share schemes combined with keys, or only keys to obtain computational security, respectively. We believe that the information theoretically secure level we obtain in SSSDB while supporting search operation efficiently using the accumulating automata approach is significantly better.

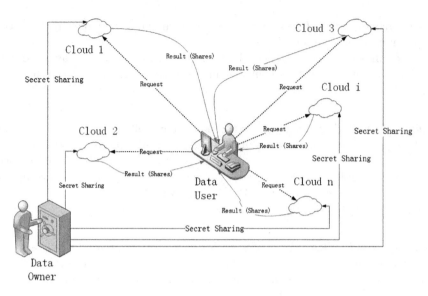

Fig. 1. Settings of SSSDB

1.2 Our Contributions

In this paper, we propose the SSS method, which allows searching and collecting statistics inside secret shared private information, without retrieving that information. Then, we utilize the SSS to create a SSSDB scheme. The framework of the SSSDB scheme is depicted in Fig. 1, where the Data Owner secretly shares their data and sends the database share to public clouds separately. The Data User sends the query share to each cloud and gets the result share from each cloud. Each public cloud works on the database share and the query share independently to calculate its result share. At the end, the Data User can reconstruct the correct result from the result shares while Clouds know neither the database and the query results. Actually, we extend the application of *Accumulating Automata* [4] by secret sharing every character in the searched pattern. Namely, we are able to secure the search pattern by sending a vector of secret shared bits for selecting each character in the searched pattern.

Online Search. In SSSDB, if the client only wants to check the presence of a certain allergy A in a specific medical record R, assuming R is associated in the database with a unique ID N, it generates a private query from N and A, and gets an indication I whether A is present in R, without retrieving R. The response, i.e., I, is also private as it is secret shared as well.

The online search has three advantages over retrieving R and searching for A in R locally:

Delegating computation: The search is not consuming the client computational resources.

Short answer: Only I is sent to the client, and not R, which reduces the communication complexity.

Concise answer: The presence of R on the client machine, may introduce privacy hazards, as if the client machine is stolen or hacked, R may be compromised.

Privacy Preserving Statics. A company may collect statistics about the database in a single query, without revealing any information on the records, except the statistics, in the form of the number of appearance of the queried string. For example, consider a database of medical records. A medical company may run a private query to count the patients who have a certain disease, and therefore predict the demand for its cure.

No Hash. Unlike TRANSPIR, SSSDB does not need a potentially leaking auxiliary access control data structure that maps keys to unique indexes in a way that reveals the current set of keys and in turn reveals information on the records in the secret shared files (in the clouds).

Dynamic Update. SSSDB is the first dynamic private database we are aware of. By cutting the connection of keys and specific indexes in the database tables, SSSDB allows clients to securely update the database tables by inserting and deleting arbitrary $\langle key, pattern \rangle$ pairs.

1.3   Organization

In Sect. 2 we establish SSS, a method for searching private information which may reside in the public cloud, without retrieving the private information. Section 3 explains how to employ SSS to build SSSDB, a dynamic database with private information, and online searches. Section 4 describes SSSDB implementation and its usage for storing (health) records. We conclude in Sect. 5.

2   Private Searchable Data

In this section we explain how SSS is working, for more detailed information please refer to *Accumulating Automata* [4]. The classic Shamir secret sharing scheme [12] allows a dealer to choose a secret value, and distribute shares of that secret to many players. For the sake of simplicity, we present our idea where each letter in the data to its unary representation, so if the alphabet α is $A_1...A_l$, then the letter A_k, $1 \leq k \leq l$, is represented by an array of size l, with 1 at index k and 0's in all other slots. Now, if a file F holds s letters, $L_1...L_s$, and the alphabet is α, it is converted to a matrix M^F of dimensions $l \times s$, where if $L_m, 1 \leq m \leq s$ is A_k, then we have 1 at $M^F[m, k]$ and 0 at $M^F[m, j]$, $j \neq k$. At this point we generate p secret shares of M^F. If σ is the length of the longest pattern we intend to search in F then, to facilitate the search method, which we explain shortly, $p = (2 \times \sigma) + 1$. $M_q^F, 1 \leq q \leq p$, is a matrix of $l \times s$ numbers, $M_q^F[i, j]$ is a secret share of $M[i, j]$.

At this point we distribute each share, M_q^F, to a separate cloud, and verify that no single entity will hold two shares of F. Note that, even if an entity maintains shares of many files, it should identify the exact shares of a specific file, it may also need to map the shares to the right x's they represent.

The preprocessing secret share construction stage described above, is quite resource consuming, both in computation and communication, but once it is done we are ready to repeat indefinitely very efficient searches in F for any pattern π that is p letters or less over α.

We convert π to its unary representation and generate the matrix M^π, and share M^π to the p hosts in the cloud (or different clouds). It should not take too much time or communication, as σ is usually small. Now the host H_q holds both M_q^F and M_q^π. Each host should calculate a number, that is the y coordinate for its x, in a polynomial of degree p. The calculation is given in the following equation:

$$(\Sigma_{k=1}^{s-\sigma}(\Pi_{i=1}^{\sigma}(\Sigma_{j=1}^{l}(M^F[k+i,j] \times M^\pi[i,j])))) \tag{1}$$

The polynomial created by the points that are calculated with Eq. 1, will have the degree $p-1$, and when x is 0, will be the number of times π appeared in F. The reason is that the inner sum, of multiplications of the components of the unary representation of a letter in α, is a polynomial of degree 2, which is 1 at $x = 0$, only if the character in π and F are the same.

Security. In SSS the clients distribute the secret shares and can interpolate them, while the servers, have no knowledge of the data. Clients are free to communicate with each other and the servers, but the servers are assumed to not communicate with each other, and are in fact, isolated from one another (possibly competing with suppliers from different clouds in different geographic areas).

Initialization Complexity. At initialization, the client is creating the secret shares of the data. If the length of the maximal search pattern is p, then there are at least $2 \times p + 1$ shares. Each share involves calculating and storing $n \times s$ values, where n is the alphabet size and s is the number of letters in the data. The communication overhead is uploading $((2 \times p + 1) \times (n \times s))$ values, and the size of a values in $\mathbb{F}$. The $\mathbb{F}$ size should be at least $\max(p, output)$. Each server needs to allocate storage for $(n \times s)$ values.

Search and Counting Query Complexity. As in initialization, the client creates and distributes $2 \times p + 1$ shares of the pattern, but now S_P, the size of the pattern, replaces S_D, and is small. Each server during the search is performing $n \times S_P \times S_D$ multiplications, and sends a value, which is the result of the calculation in Eq. 1 to the client. The client then needs to interpolate the polynomial to decode the results.

3   SSSDB Structure

We leverage SSS to create SSSDB, a database that allows insertion and deletion of $\langle key, record \rangle$ pairs. SSSDB can be used to retrieve records privately according to keys, or to search for a pattern in a private online record, according to a given $\langle key, pattern \rangle$ pair. SSSDB uses the keys directly as pointers to the data tables, which avoids the need for maintaining an auxiliary access data structure outside the secret shared database.

In Sect. 3.1 we present the objects that add up to create SSSDB and in Sect. 3.2 we discuss the details of SSSDB operations, namely, retrieving and online search, insert and delete.

3.1   Database Components

The database always holds n $\langle key, record \rangle$ pairs, where each key size is u and each record size is r. n is the actual maximal number of keys expected in the database, and not the keys universe, i.e., not the maximal possible number of different keys in the system.

We define the special characters $\emptyset$ and $\perp$, which will never be used in the client keys and records, and use $\perp$ for keys in empty slots and $\emptyset$ for their records. If there are u empty slots in the database, they will be marked $\perp_1....\perp_u$. $\perp_v$ is a unique deleted key starting with the character $\perp$ that is followed be characters that encode the sequence number of the leftover (deleted) entries.

In initialization, the servers allocate an array of key-record pairs to accommodate n $\langle key_j, record_j \rangle$ pairs, where n is the maximal actual data size. The initializing client then secret shares $\langle \perp_j, \emptyset's \rangle$ to each empty slot. We emphasize that not all slots must be empty in initialization, which is important in preventing an eavesdropper from distinguishing an insert from a delete.

We set a counter C, which is shared with all clients, to n, to indicate there are n free slots in the database, which are marked $1...n$. If there are multiple clients, C must be adjusted to maintain serialization of the database operations, therefore in this paper we consider, for the sake of simplicity (as for synchronization and locking techniques during updates), the existence of one server that serves all clients.

SSSDB defines a simple application protocol, which allows the client to call commands on the server side. When we list the steps of a database operation, if the step is executed on the host, it is executed as a result of a request sent with this protocol, and it sends an output or acknowledgement to the client.

3.2   Database Operations

Now we show how the database components and SSS are employed to create the operations of our SSSDB.

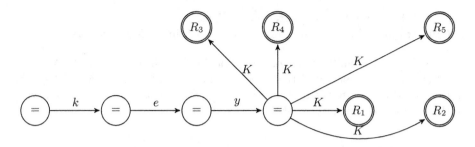

Fig. 2. Extracting the record R, of size 5, with key K

Retrieving Search. A search operation executes the following steps:

1. **Client:** Gets a valid key K of size k. It generates and distributes the SSS secret shares for K.
2. **Each Server:** Uses SSS to generate the record of K, if it exists, or a record of all 0's. Then each server sends the results to the client.
3. **Client:** Interpolate the record from the servers.

The search retrieves the record for a key, without knowing the index.

Step 1 is common to all operations where it creates and distributes the shares of M^K. After completing step 1, the server's side, is calculating R in step 2. The process of calculating R for a 5 letters record $\langle R_1, ..., R_5 \rangle$, is illustrated in Fig. 2. It starts with an SSS online search for the pattern, e.g. $\langle k, e, y, K \rangle$ from Fig. 2. As the table is populated with n $\langle key, record \rangle$ pairs, and all keys are of size k and all records are of size r, there are exactly n sequences of k characters that may accommodate the key $K_1, ..., K_k$. For each such sequence, S^j, $j \in 1...n$, we sum the $K_i \times S_i^j$, $i \in 1...k$, to get Q^j which is 1 if S^j is K and 0 otherwise. The server now multiplies Q^j by R_i^j, $i \in 1...n$, to get n values. These values are R_i^j if Q^j is 1 and 0 otherwise. The function sums $R_i^j \times Q^j$ for $i \in 1...n$, to get r secret shared values. As K may appear in an SSSDB table at most once, these values are the record for K, if K is found, or 0's. The server sends r results to the client which is then constructing the record.

Non Retrieving Counting in Records. In this operation the servers get a pattern π and a key K_j and calculates how many times π appears in record R_j which is associated with K_j. The phases are as follows:

1. **Client:** Gets a valid key K_j and a pattern π, generates their SSS shares and distributes to the servers.
2. **Each Server:** We break this step to the following:
   (a) Use SSS to generate the share of record R_j, which is associated with K_j, as done in Sect. 3.2.
   (b) Uses Eq. 1 with s set to size of R_j, to compute the secret share of the number of appearances of π in R_j, if K_j exists in the database, or 0 otherwise.

(c) Send the result secret share to the client.
3. **Client:** Interpolate the appearances count of π in R_j from the servers.

This function is built on top of the retrieving search from Sect. 3.2. At step 2c, after the server generated the r secret shares, $R_1...R_r$, which are either 0's or R, it goes to execute the SSS search on $R_1...R_r$. If R is found, the result will be the number of appearances of π at R, and otherwise, it would be zero.

Global Statistics. SSSDB has a special command to count a pattern in the whole database. As the keys contain the $\perp$, and the records are initialized to $\emptyset$s when deleted, it is possible to count the total number of appearances of a pattern in the database. This can be used, for example, to generate useful statistics such as spread of epidemics by counting a disease name in online private medical records.

Index. Both insert and delete require knowing whether a key K is in the database and what is its index. For this purpose, we have the function $index(K)$, which is working like the retrieving search from Sect. 3.2, but it substitutes the index, $1...n$, instead of R. The index exists anyway in the process of server iteration, and is not secret shared. $index(K)$ allows the client to calculate the index of K which is $1...n$ if K is in the database or 0 if it is not.

Delete. In this section and in Sect. 3.2 we assume the existence of $index(K)$ function from Sect. 3.2, and ignore, for simplicity of presentation, the fact it executes on the host. The delete sequence:

1. **Client:** This step goes through the following phases:
   (a) Get a valid key K.
   (b) Calculate the index X of K, from the servers, and if X is 0 (K is not in the database), terminate.
   (c) Fetch C into Y and increment C.
   (d) Generate SSS shares for $\perp_Y$, which is an unused key for an empty, deleted slot. The process is described in Fig. 3.
   (e) Distribute the shares of $\langle \perp_Y, 0's \rangle$ to the server, together with plan X.
2. **Each Server:** replace the $\langle key, value \rangle$ in index X with $\langle \perp_Y, 0's \rangle$, as demonstrated in Fig. 4.

Step 1a is checking for the index of K. If K is absent, the delete operation terminates, as there is nothing to delete. In step 1c the client uses C. C is a counter of deleted keys, and deleting a key K means incrementing C and then replacing K with a key that is a combination of $\perp$ and the current C, i.e., $\perp C$. $\perp C$ must be unique, because $\perp$ is never used in client keys, and C will be decremented only if an insert used $\perp C$ to insert a $\langle key, record \rangle$ pair.

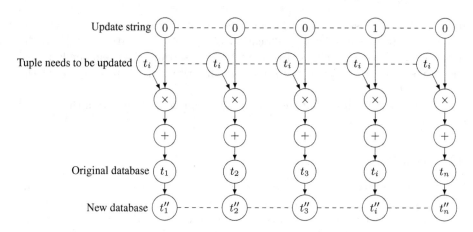

Fig. 3. Create the refreshing data

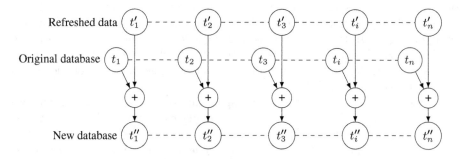

Fig. 4. Update algorithm on secret shared database

Insert. The insert sequence is as follows:

1. **Client:** This step goes through the following phases:
   (a) Get a valid key K, and a record R.
   (b) Calculate the index X, of K from the servers, and if X is not 0 (K exists in the database), terminate.
   (c) Fetch C into Z and decrement C, and calculate the SSS secret shares for $\perp_Z$, a used key, which marks an empty slot.
   (d) Calculate the index Y of $\perp_Z$, i.e., the index of an empty slot in the database table.
   (e) Generate SSS shares for $\langle K, R \rangle$ as well, and distributes it together with plan Y. The process is described in Fig. 3.
2. **Each Server:** Replace the pair in index Y with $\langle K, R \rangle$, as illustrated in Fig. 4.

Step 1a is checking for the index of the new key, K. If it is present, the insert operation terminates. Now we need to find an empty slot for the insertion of $\langle K, R \rangle$. For this, we allocate a key which we know is occupying an empty slot, namely, $\perp_C$. We assume there is only one client in the system, otherwise we would have to synchronize the allocation of $\perp_C$. At step 1d we set Y to the index of the empty slot of $\perp_C$. To complete the insertion, the client sends Y and the SSS shares of $\langle K, R \rangle$ to the server which writes $\langle K, R \rangle$ to index Y of the table.

Now it is left to show that the operations we presented, maintain the privacy of the operations.

3.3 Security

The search security is given by SSS security, which in turn is as secure as PIR. The more challenging part is to show that privacy is kept in the dynamic updating operations. Cloud servers can track the updates of entries, but this information, as we explain below, only reveals the time, possibly measured in the number of updates elapsed since the last update of each entry in SSSDB. However, if one wants to avoid such leakage, it is possible to refresh by adding secret shared zeros to all but the updated entry. We can also somewhat hide the exact access by updating (refreshing) a subset of other entries in addition to the entry we need to update, rather than refreshing all. We recall that refreshing the secret shares is always beneficial, as if an eavesdropper copied a subset of the shares they will become useless.

To demonstrate the security advantage of inserts and deletes in SSSDB, we assume two things. One is that the eavesdropper can not decode secret shares, which is a fact. The other is that the dealer/inquiry server is trusted. The client knows the keys and records, as it is the one generating them, thus we must trust the client.

To show that any key can be in any index, thus knowing the index, and leaking less information about the data, we will present a scenario where two full keys are swapping indexes and a scenario where two empty keys swap indexes. The scenario where any empty key may replace any full key, and the scenario any full key may replace any empty key are trivial. We also assume that the servers can not tell whether the table is full or empty, thus the servers can never know if the update it intercepted was an insert or a delete.

Swap Full Keys. Assume $n = 10$ and $C = 5$. The client deletes K_i and replaces it by $\perp 6$ and increments C to 6. Then it deletes K_j, $j \neq i$, replaces it by $\perp 7$, and increments C to 7. Now the client inserts K_i, which goes to $\perp 7$ which is in the previous index of K_j. To complete the swap, it inserts K_j, which goes to $\perp 6$ which is in the previous index of K_i.

Swap Empty Keys. Assume $n = 10$ and $C = 5$. The client inserts K_i and replacing $\perp 5$ and decrements C to 4. Then it inserts K_j, $j \neq i$, replacing $\perp 4$, and increments C to 3. Now the client deletes K_i, which is replaced by $\perp 4$. Recall K_i replaced $\perp 5$, so $\perp 4$ swapped with $\perp 5$.

4   Evaluation

To check how our private database actually works we implemented it in Python 2.7, using standard libraries. We choose EC2 to store our shares and use sample US army veterans medical records as our testing data.

Testing environment: We used EC2 smallest VM, i.e. T2 micro, running Amazon Linux AMI, for the database servers, and our local machine, Intel Core I3-2310 M, 2.1 GHz, with 4 GB RAM, running Windows 7 Home Premium, which is a standard laptop, as database clients.

Table 1. Pattern and data sizes are in bytes, while time for initialization an operations is measured in seconds.

Global statistics			
Data	Pattern	Initialization	Operation
4 K	5	79	5
4 K	10	79 (reuse)	8
8 K	5	163	8
8 K	10	163 (reuse)	14

In Table 1 we show the results of global statistics collections. We count the occurrences of 5 and 10 bytes patterns in 4 KB and 8 KB files. It can be seen the initializations are long, but then extracting the statistics takes only a few seconds.

Table 2. Keys and record sizes are in bytes, while time for initialization an operations is measured in seconds.

Retrieving search				
Keys #	Key	Record	Initialization	Operation
10	5	4 K	23	133
100	5	400	45	17
1000	5	40	83	9
10000	5	4	220	8

Table 2 is uploading a database, whose plain text is 40 KB, and is configured to hold records of different sizes. From 1000 records of 4 bytes to 10 record of 4 KB. The keys size is 5 bytes. In the retrieving search operations, the bottleneck is the calculation in the interpolation of the record, and in the initialization, it is the calculations in creating the keys which is in unary representation. It can be seen that the interpolation is much more consuming. We note that we have started a new implementation using c++ indicating the potential for a dramatically better performance.

5   Discussion

This paper presents SSS, a method to search in private information without retrieving it, in an efficient way, and SSSDB, a database based on SSS. In SSS the clients distribute the secret shares and can interpolate them, while the servers have no knowledge of the data. Clients are free to communicate with each other and to servers, but the servers are assumed not to communicate with each other and are, in fact, isolated from one another (possibly in different clouds in different geographic areas).

Search security is given by SSS security. The more challenging part is to show that privacy is kept in the dynamic updating operations. Cloud servers can track the updates of entries, but this information only reveals the time, possibly measured in number of updates elapsed since the last update of each entry in SSSDB. However, if one wants to avoid such leakage, it is possible to refresh by adding secret shared zeros to all but the updated entry. We can also somewhat hide the exact access by updating (refreshing) a subset of other entries in addition to the entry we need to update, rather than refreshing all.

The complexity of all operations involves a search which takes $2 \times \log^2 n + \log n$ communication. The non retrieving operations from Sect. 3.2 does not add this basic complexity, whereas the other does:

1. **Retrieving Search (Sect. 3.2):** Adds the size of a record from each server.
2. **Delete (Sect. 3.2):** A delete operation is performing only one search, to find the index of the to be deleted key, and then replaces the full key with a deleted key, so it sends another key, a secret shared empty key. Assuming the key size is $\log n$, we get $4 \times \log^2 n + \log n$.
3. **Insert (Sect. 3.2):** An insert operation may execute two searches. The first, to find that the key is absent, the second to find the index of the empty key which it allocated. It then also sends the secret shared $\langle key, record \rangle$ pair, so the communication complexity, again assuming key size is $\log n$, is the size of the record plus $6 \times \log^2 n$.

On our weak laptop, with no optimized python implementation, and the slowest EC2 servers, a search for a keyword in a medical record took only a few seconds. In addition to searching online, SSSDB allows dynamic updates and does not use auxiliary access data structure. We note that standard methods such as Berlekamp Welch techniques may be used to interact with servers, where some of which are malicious.

Our future work includes: (1) How to securely store more than one database share in one public cloud, say, EC2; (2) How to implement a flat database or a Mongodb-like database based with the SSSDB idea.

References

1. Security First Corp. https://www.securityfirstcorp.com/, https://www.linkedin.com/company/navajo-systems
2. Agrawal, D., El Abbadi, A., Emekci, F., Metwally, A., Wang, S.: Secure data management service on cloud computing infrastructures. In: Agrawal, D., Candan, K.S., Li, W.-S. (eds.) Information and Software as Services. LNBIP, vol. 74, pp. 57–80. Springer, Heidelberg (2011)
3. Chor, B., Kushilevitz, E., Goldreich, O., Sudan, M.: Private information retrieval. J. ACM **45**(6), 965–981 (1998)
4. Dolev, S., Gilboa, N., Li, X.: Accumulating automata and cascaded equations automata for communicationless information theoretically secure multi-party computation. Cryptology ePrint Archive, Report /611 (2014). http://eprint.iacr.org/
5. Gentry, C.: Fully homomorphic encryption using ideal lattices. In: Mitzenmacher, M. (ed) STOC 2009, Bethesda, MD, USA, May 31–June 2, 2009, pp. 169–178. ACM (2009)
6. Gentry, C.: Toward basing fully homomorphic encryption on worst-case hardness. In: Rabin, T. (ed.) CRYPTO 2010. LNCS, vol. 6223, pp. 116–137. Springer, Heidelberg (2010)
7. Gentry, C., Halevi, S.: Fully homomorphic encryption without squashing using depth-3 arithmetic circuits. In: FOCS 2011, pp. 107–109. IEEE Computer Society (2011)
8. Gentry, C., Halevi, S.: Implementing gentry's fully-homomorphic encryption scheme. In: Paterson, K.G. (ed.) EUROCRYPT 2011. LNCS, vol. 6632, pp. 129–148. Springer, Heidelberg (2011)
9. Hadavi, M.A., Jalili, R.: Secure data outsourcing based on threshold secret sharing; towards a more practical solution. In: Proceedings VLDB, Ph.D, Workshop, pp. 54–59 (2010)
10. Liu, Y., Wu, H.-L., Chang, C.-C.: A fast and secure scheme for data outsourcing in the cloud. KSII Trans. Internet Inf. Syst. (TIIS) **8**(8), 2708–2721 (2014)
11. Popa, R.A., Redfield, C.M.S., Zeldovich, N., Balakrishnan, H.: CryptDB: protecting confidentiality with encrypted query processing. In: Proceedings of the Twenty-Third ACM Symposium on Operating Systems Principles, SOSP 2011, pp. 85–100. ACM, New York (2011)
12. Shamir, A.: How to share a secret. Commun. ACM **22**(11), 612–613 (1979)
13. Stefanov, E., Shi, E., Song, D.: Towards practical oblivious RAM (2011). arxiv:1106.3652
14. Tian, X.X., Sha, C.F., Wang, X.L., Zhou, A.Y.: Privacy preserving query processing on secret share based data storage. In: Yu, J.X., Kim, M.H., Unland, R. (eds.) DASFAA 2011, Part I. LNCS, vol. 6587, pp. 108–122. Springer, Heidelberg (2011)

Graph DBs vs. Column-Oriented Stores: A Pure Performance Comparison

Marios Kendea[1]($\boxtimes$), Vassiliki Gkantouna[1], Angeliki Rapti[1], Spyros Sioutas[2], Giannis Tzimas[3], and Dimitrios Tsolis[4]

[1] Computer Engineering and Informatics Department,
University of Patras, 26504 Patras, Greece
{kendea,gkantoun,arapti}@ceid.upatras.gr
[2] Department of Informatics, Ionian University, 49100 Corfu, Greece
sioutas@ionio.gr
[3] Computer and Informatics Engineering Department,
Technological Educational Institute of Western Greece, 26334 Patras, Greece
tzimas@cti.gr
[4] Department of Cultural Heritage, Management and New Technologies,
University of Patras, 26504 Patras, Greece
dtsolis@upatras.gr

Abstract. Cloud Computing has brought a great change in the way information is stored and applications run. In order for one or more clusters to work as a cloud we need a middleware framework, such as Apache Hadoop [17], that provides reliability, scalability and distributed computing. Once the infrastructure has been established, a software framework can be installed, which runs on top of it and will be the connection to communicate with the applications developed by the users. The software, in this regard, is a NoSQL database. This paper deals with the problem of searching data in some widespread NoSQL databases used in cloud computing. Two categories of NoSQL databases are compared; one based on columns using a column-oriented key-value store, HBase [6], and a high-available graph database, Neo4j [11]. HBase is a distributed, scalable storage system that runs on top of HDFS, and has being designed based on Google's BigTable [4]. Neo4j has being designed and developed to be a reliable database, optimized for graph structures, instead of tables, and is a robust, scalable, high performance and high available database that supports ACID transactions and queries written in Cypher language. The aim of this paper is to create a novel system that will decide when a query must be send to be executed in a key-value store or a graph database. Thus, an experimental pure performance comparison has been made between Apache HBase and Neo4j for a variety of queries, that were programmed using systems API's and Java language.

Keywords: NoSQL databases · Hadoop · HBase · Neo4j · Graph databases · Distributed systems

© Springer International Publishing Switzerland 2016
I. Karydis et al. (Eds.): ALGOCLOUD 2015, LNCS 9511, pp. 62–74, 2016.
DOI: 10.1007/978-3-319-29919-8_5

1   Introduction

The massive data generation and the need of real time query execution on them lead the sql-like databases to be less useful. Furthermore, relational databases can store large amounts of structured data with high performance and of course ACID transactions. Because of the data explosion and the turnover to cloud technologies, a new era of databases is born. That's the basic reason of the existence and development of the wide range of different types of data stores, called NoSQL databases. This data may not follow a specific schema and the NoSQL databases give us the chance of not worrying how we will explicit use the raw data.

NoSQL databases have gain great attention consisting different types. The most popular are Document databases, Key-Value stores, Column-oriented and Graph stores. Some non-structure data storage NoSQL systems are HBase [6], MongoDB [9], Dynamo [5], Cassandra [10], and Neo4J [11] etc. In this paper, we are considering two of these categories and one open source system for each one respectively. The first system is the Column-Oriented HBase and the second one is a Graph store called Neo4j. Columnn-Oriented stores save the data within columns in tables and each row is accessible via the key. Tables can be sparse because each line can have different attributes than the other. A graph database stores data in a graph, the most generic of data structures, capable of elegantly representing any kind of data in a highly accessible way.

In order to create modern applications and satisfy users we need to test these systems. Lizhi Cai et al. in [3] present a performance testing model for those complicated systems pointing that performance testing should start from two aspects: the architecture and business level. We follow up the first/second level and we contribute an evaluation on the systems above.

The rest of this paper is organized as follows. Section 2 presents an overview of the frameworks used. Section 3 then describes the dataset used and the types of queries used to evaluate the systems. Section 4 discusses our results and finally in Sect. 5 we conclude the paper.

2   Motivation

In a research area full of new NoSQL databases, we would like to see how a pure performance comparison, between the two most popular types, Column-Oriented and Graph Databases, can give us results that can be used to create a model to distribute different kind of queries between databases in Cloud applications. Different work is being made on this area, like comparison between a Graph and a Relation database [16], but not a comparison between a Graph and a Column-Oriented stores. A large part of publications about Graph Databases is being evaluated on Neo4j Graph store [11], the different ways that this open-source framework can be used for query processing [18], the feature of indexing and that is 2–5 times faster than MySQL [7] are some of the reasons that led us to this choice.

We made the choice to compare two systems with different data model. For our data, Graph Database is the clear first choice, because we are interested in graph data, but what we are trying to do is to see if only the data can led us to the right choice and how a combination with an other system might give us flexibility for more complex and critical applications.

3   Systems Overview

3.1   NoSQL Features

There are some characteristics that the large-scale distributed systems have to implement at the minimum. Those are security, scalability, high fault-tolerance and high- availability. Unfortunately, this characteristics can't meet at the same time. In [2] an updated version of CAP theorem is presented, which is a good description about that fact. Figure 1 shows the distribution of some systems and hence their lack in characteristics. The new rules say that we can achieve some trade-off of all three. For example, HBase to have data consistency and partition tolerance has some loss in availability.

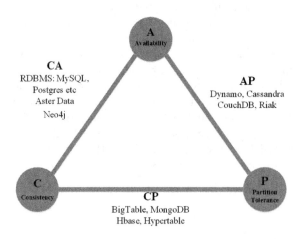

Fig. 1. The NoSQL systems under the CAP theory

3.2   HBase

HBase is an open-source implementation of Bigtable [4], which runs on top of Hadoop framework [17] that allows distributed processing. The logic behind HBase is that tables can be created without any specific schema, and all columns of a row can be accessed by a key. There is no support of queries and so there is no direct access on columns of a row. Table 1 shows an abstract example of a

Table 1. Column-oriented storage example

Row key	Time stamp	ColumnFamily1		CF2	CF3
		CF1-q1	CF1-q2	CF2-q2	CF3-q3
rk1	t1	value1			
rk2	t2		value2		
rk3	t3			v3	
rk4	t4				v4
	t5				v5

column-oriented storage. One cell of the table can be determined by the three properties: row key, the timestamp, and column. Row key is the unique identifier for the rows, timestamp is the data version assigned by the system. The column is defined as the form "column family: qualifier". A table can have multiple column families, one column family is constituted by multiple columns.

HBase stores data in bytes form, alphabetical ordered and always returns the line of data with the biggest timestamp. HBase provides distributed storage and data access. Dividing the set of keys of a table in separate parts and assigning each part to a separate cluster node-*RegionServer*, HBase ensures that the load balancing of the system nodes is as evenly as possible. The Hadoop Distributed File System (HDFS) [14] replicates table parts to ensure fault-tolerance.

3.3   Neo4j

Well known companies in the world like Google, Facebook and Twitter have a common between them and that is they have connected data as the center of their plans. Neo4j is the implementation chosen for this paper, on previously unknown type of Graph No-SQL databases. It is open source for noncommercial use. Quickly became one of the most popular graph database systems. According to the Neo4j website, "Neo4j stores data in nodes connected by directed, typed relationships with properties on both, also known as a Property Graph" [11].

The main features are the following [13]:
- it uses a graph model for data representation
- it's reliable, with full ACID transactions
- it's custom disk-based, durable and fast
- it has native storage engine
- it's massively scalable, up to several billion nodes/ relationships/ properties
- it's highly-available, when distributed across multiple machines,
- it's expressive, with a powerful, human readable graph query language (Cypher),
- it's fast, with a powerful traversal framework for high-speed graph queries and embeddable and accessible by a convenient REST interface or an object-oriented Java API.

Neo4j supports indexing on specific attributes of nodes so the queries can be written and executed easily and of course spending a small amount of time. Imagine if you had to traverse a giant graph to find the nodes containing a specific attribute.

Graph Data Model

Graph database models use data structures modeled as graphs or generalizations of them [1,8]. The property graph model has the following characteristics:

- Let $G = (V, E)$ be a directed graph where V is a set whose elements are called vertices or nodes, and E a set of ordered pairs of vertices, called arcs, directed edges, or arrows. In this case V can be a set of sets of nodes of a particular type defined by the data the graph represents. $V = \{V_1, V_2, ..., V_{n1}\}$ where $V_i = \{v_{i1}, v_{i2}, ..., v_{in_2}\}$. The same applies for the edges of a graph.
- Each node can contain attributes.
- Relations can have a type, are always directed and always connect two nodes.
- Each relation can contains attributes.

4   Design of Experiments

For our experiments we established a 3 computer nodes cluster running with Hadoop-1.0.4 and HBase-0.94.5 which are connected with public IP's, the configuration is: CPU: 4 cores, memory: 8GB, disk: 40GB, os: 64bit Debian. The same cluster used for Neo4j evaluation running in High Availability mode (HA) using the version neo4j-1.9.4.

4.1   Dataset

We have created our own dataset, which can be found online[1], by collecting movies data including persons data which participate with different roles infront or behind the scenes. Movie data contain different attributes, not the same for each movie so we are not violating the Hbase schema-less design. The dataset contains about 205.000 movies and about 21.000 person information. Furthermore, we have the relations between movies and persons and the different types that a movie is categorized. Figure 2 shows the difference between the two systems in storage terms for the same data described in this subsection and based on the design choices made described in the following. We can see that the chosen HBase schema gets about 2 times less space than Neo4j Store.

---

[1] https://github.com/kendea/dataset_movies.

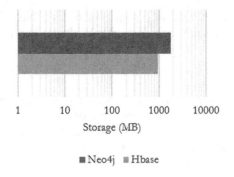

Storage (MB)

■ Neo4j   ■ Hbase

Fig. 2. Occupied database storage for the same data

4.2   Design Choices

HBase doesn't support indexing on data so we had to find a way to associate the data of the different tables in an efficient way. We created some extra HBase tables from the data, whose basic structure is like an inverse index. For different data we created different indexes so we can measure later how this different indexes affect the time of queries execution. At this point, is important to highlight the fact that we are dealing with real-time execution, so every aspect that is possible to reduce the run time must be taken into account. First, we tokenize the text of movie_titles or movie_synopsis and then we create an inverted index on produced (tokenized) words to columns of movie_ids. An other index table contains movie genres of specific type, as keys. Thus, we had to create very long lines to store the movies ids in HBase and examine how the big number of columns could affect the overall query performance. We split the HBase tables so every part goes to a different cluster node. As mentioned earlier, we are interested in real-time processing so we couldn't use Apaches' Hive [15], which in some point uses inverted index as we do, but it has been implemented for batch analysis to large data sets, so it was not suitable for our use case. A lot publications suggest different types of multidimensional indexes running on top of HBase for efficient query execution, but most of them are used for location aware services [12] where the data can be represented in mathematical form. We choose to use inverted indexes as the simplest implementation and by this way we can generalize our model in different kind of datasets. The idea of inverted index is coming along the naturally way of thinking the key-value stores capabilities.

Neo4j supports indexing on specific properties for a specific type of nodes. In that in mind, it's easier to create a "schema" that follows the data used. We created nodes for movies, person and genres all indexed. With this choice we can see how "fat" nodes affected positively or negatively varying difficulty queries. By the term fat referring to a node means that this node has a large number of in-direct or out-direct nodes. The relations are of two different types. The first type connects a person with a movie while the second connects a movie with the nodes representing the different genres of movies. Neo4j runs in HA (High Availability) mode meaning that every node in cluster contains the same graph.

4.3   Queries

We have contacted different scenarios of queries that might be asked to a movie database. Of course we had a more open minded thought when creating those scenarios so as in case the data representing something different, to have a generalization on our results. We programmed a variety of query types from simple ones to more complex ones and we executed a very large number from each category on both systems running on a single cluster. Below, we describe the queries logic and provide query examples. HBase doesn't support any language for queries so we programmed them manually using the JAVA API. Neo4j supports Cypher language having the ability to be able to give automatically the different arguments of the queries. In some cases we use JAVA API provided by Neo4j to execute the simple queries. The source code and the query engine can be found online[2].

Simple Aggregate Queries

Q1: These queries calculate and return the distinct cardinality of objects on a specific attribute with different number of attribute values using AND or OR operators.

> *COUNT the movies WHERE contain the word "word1" in title or synopsis*
> *COUNT the movies WHERE contain the words ⟨ "word1" AND "word2"⟩ in title*
> *COUNT the movies WHERE contain the words ⟨ "word1" OR "word2" OR "word3"⟩ in synopsis*

Q2: This query uses the same logic as above but this time returns also the records obeying the logical expression between attribute values AND. In this case we use an inverted index to find the movies ids and then we "join" this with the HBase table containing the information for that movies. Joins are not supported so we manually programmed them. In particular they are partly joins because we know which lines are retrieved by the keys resulting at the first step of join.

> *FIND the movies WHERE contain the words ⟨ "word1" AND "word2" AND ... AND "wordn"⟩ in title or synopsis and RETURN the information stored*

Q3: This query is the same as Q2 but this time we are sure that the attribute values apply for the operator AND. This query is important for applications that might use data with auto-filling and do not allow custom text from user.

---

[2] https://github.com/kendea/hbase_neo4j.

> *FIND the movies WHERE contain the words ⟨ "word1", "word2"⟩ in title and RETURN the information stored*

Range Queries

Q4-Q5: Both of these queries are executed on one attribute. The first query (Q4) returns records lexicographically close to the attribute values of the query. This type of query is popular among applications that user might misspelled a word or didn't type the whole thing. The second query (Q5) returns only the records that are matched exactly. Both of this queries join 3 tables in HBase to produce the results. The joins are programmed optimally so each previous step of join produces less results than the next one.

> *(Q4) FIND the people WHERE their name looks like the text "name" and RETURN the information stored for them and their movies*
>
> *(Q5) FIND the people WHERE their name contains the text "name" and RETURN the information stored for them and their movies*

The following 2 queries are more complex.

Q6: This query gets arguments for two attributes and involves 2 HBase tables; so 1 join. Neo4j query has to find the nodes from index that apply for the first attribute and the same time look inside those nodes for the second attribute.

> *FIND the movies WHERE contain the words ⟨ "word1" AND "word2"⟩ and their year production is in the range [year1,year2] and RETURN the information stored*

Q7: This query gets arguments for three attributes and involves 2 HBase tables so 1 join. As before, Neo4j uses the index and then look inside the node for the rest information.

> *FIND the movies WHERE contain the words ⟨ "word1" AND "word2"⟩, their year production is in the range [year1,year2] and the movies genres must be [genre1, genre2, ...] and RETURN the information stored*

Group-By Queries

Those queries are basically the same executed on different attributes to see how the design of HBase index tables and Neo4j "fat" nodes affect this type of queries.

Q8: This query groups results on one attribute in a range of values. This attribute is not indexed in any way on both HBase or Neo4j.

> *FIND the movies WHERE contain the words ⟨ "word1" AND "word2"⟩, the year production is in the range [year1,year2] and the movies genres must be [genre1, genre2, ...] and GROUP BY year*

Q9: This case groups results on one attribute with arguments, if there are more than one, connected using operator *AND*. This time, we used HBase index table and "fat" nodes for each object of this attribute.

> *FIND the movies WHERE contain the words ⟨ "word1" AND "word2"⟩, the year production is in the range [year1,year2] and the movies genres must be [genre1, genre2, ...] and GROUP BY genre*

Q10: This query is even more complex because is a combination of both of previous queries (Q8 and Q9).

> *FIND the movies WHERE contain the words ⟨ "word1" AND "word2"⟩, the year production is in the range [year1,year2] and the movies genres must be one of [genre1, genre2, ...] and GROUP BY year and genre*

Top-K Queries

Q11: The last query examined was top-k. We used the most complex of query from the previous, but this time we are not returning all results. So we had to sort the nominated results on a specific attribute to see how this affects both systems. We run the experiments on random values of k.

> *FIND the movies WHERE contain the words ⟨ "word1" AND "word2"⟩, the year production is in the range [year1,year2] and the movies genres must be one of [genre1, genre2, ...] and RETURN the TOP-K GROUP BY year and genre*

Sorting

HBase returns results sorted based on key of table used. So we have to sort the results manually if that is required; which is both time-consuming programming and during execution. Cypher query language, used by Neo4j, can sort results while retrieving the data. This is the reason we executed some of the previous queries (Q1 to Q4) to see if Neo4j can sort data efficiently.

Q1-Q4: Those queries are the same as the first four presented queries but this time we change the queries to sort results on a specific property.

> *FIND the people WHERE their name looks like the text "name" and RETURN the information stored for them and their movies SORT BY name*

4.4   Analysis

We created experimental scripts for each query described in previous subsection. Each script contains a very large number of queries which are executed on both systems. This method is similar to bulk loading but instead of loading data we execute predefined queries. So the results described in the next section show the latency; which is the average time needed to complete the execution of a query. Furthermore, the analysis made corresponds to the Amortized Analysis. Amortized Complexity says that considering a sequence of operations of the same type of a program, allows the establishment of a worst-case bound for the performance of a program. In our case, a program corresponds to a query. This analysis applies when the each execution time is different and there is no clue about the real complexity of an algorithm. Amortized time can be estimated using $A_t = \frac{\sum_{i=1}^{n} T(q_i)}{n}$, where $T(q_i)$ is the time needed to execute a query of a specific time and n is the total number of queries executed.

5   Results Discussion

Figure 3 shows, in logarithmic scale of thousand of milliseconds, the mean execution time of every query in both systems. It's obvious that for the first category of queries HBase always outperforms Neo4j; but as the queries become more complex we observe that there is a turnover between the two systems.

Both systems, as we observe, in Fig. 3a drop their run time when the search is more specific (AND operator). When HBase is asked for specific keys and the result arises from the columns without more information; HBase outperforms Neo4j by far. We can also see from Q2 that partial "joins" on HBase tables doesn't affect the run time in a wide scale. Q3 performance is slightly better on Neo4j compared to the previous queries executed on the same system and that's because Neo4j can compare two strings faster instead of finding partial string similarity; when an exact match on properties of nodes is done.

The first two queries shown in Fig. 3b show us the difference in execution time of queries returning the lexicographical close results (Q4) and exact match (Q5) on one attribute. We can clearly see that HBase needs one time of magnitude larger between the two queries while Neo4j outperforms both of them. The large run time of HBase for Q4 arises from the fact that this query searches a larger space (2 joins) between multiple parts of the HBase table because of the partitioning, made by the system, on the different computer nodes of the cluster. The other two queries, Q6 and Q7, are a bit more complex because they get arguments for two and three attributes respectively. In HBase, only one join, on both of them occurs, while in Neo4j we use the built-in indexing and properties inside the nodes. We can see that Neo4j needs more time but while increasing the number of attributes involved HBase time is getting very close to Neo4j's time.

Figures 3c and 3d show the performance of the most complex and most intensive data processings. All the queries are pretty time consuming. The first two

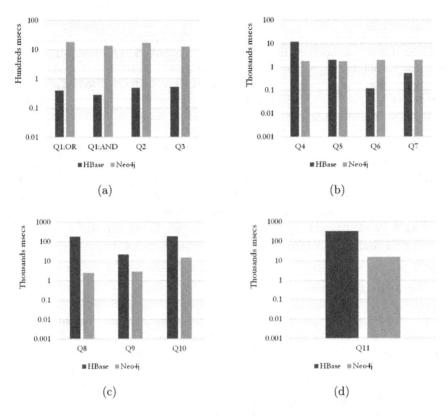

Fig. 3. Comparison for each type of query in logarithmic scale

queries, Q8 and Q9, group results on one attribute with the difference, shown in execution performance, that we used indexes. Both queries in Neo4j need one time of magnitude less, from Q9. We can now see the clear excellence of the Graph database in this kind of queries. So, we can surely say, that "fat" nodes are not problem for the data schema and they can be used freely if they can help us. For Q10, which executes a group-by on two attributes, Neo4j still outperforms HBase, but is the first time that needs more time to execute the query. For the last category, top-K queries (Q11), we can see that Neo4j outperforms HBase and that HBase needs an amount of time to sort the data, on the selected attribute, larger than Neo4j. Cypher query language supports sort while in HBase it had to be done programmatically.

The last outcome lead us to see how fast Neo4j can sort data. So we ran the first four queries (Q1 to Q4) but this time with option sort enabled. The results are shown in Fig. 4. The time needed for sorting what almost nothing. Neo4j sorts data while retrieving them.

It's obvious that in some cases HBase runs extremely better but while the difficulty and complexity of queries grows Neo4j starts to outperform HBase. Furthermore, it's easy to see that Neo4j has a steady performance for a large

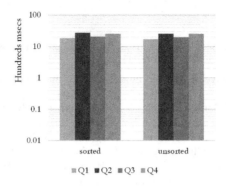

Fig. 4. Sorting using Cypher query language

number of different queries executed. The fact that in HBase we had to manually write the whole thing to debug and execute the queries was not an easy process. But based on results this is worth doing for simple queries and using at the same time different index structures based on these results. Neo4j can execute different types of complex queries very fast using indexing. Based on the results presented above, we implemented a useful and efficient query engine that distributes the queries, that are submitted by an application, to the best database in terms of pre-known performance for a selected query. In the majority of cloud use cases, the scalable cloud object storage provided from the huge volumes of racks makes the cost of replicating an application's data on two separate database systems negligible. Thus, the proposed decision making engine is of great practical interest.

6   Conclusions

In the context of this work, we presented the results on executing different types of queries, scaling in complexity, in two NoSQL databases; so we can see the performance difference between a Graph store and a Column-Oriented store with different data design choices. We have seen that we can have a distinct classification, depending on the complexity of a query and that lead us to the conjecture that it might be good for some applications to store their data in different NoSQL systems, based on the queries executed on them and an engine can make the choice based on some analysis where the query must be send to be executed.

Acknowledgments. Our thanks to C. Caratheodory Research Program from University of Patras, Greece to support this research.

References

1. Angles, R., Gutierrez, C.: Survey of graph database models. ACM Comput. Surv. **40**(1), 1:1–1:39 (2008)
2. Brewer, E.: Cap twelve years later: how the "rules" have changed. Computer **45**(2), 23–29 (2012)
3. Cai, L., Huang, S., Chen, L., Zheng, Y.: Performance analysis and testing of hbase based on its architecture. In: 2013 IEEE/ACIS 12th International Conference on Computer and Information Science (ICIS), pp. 353–358, June 2013
4. Chang, F., Dean, J., Ghemawat, S., Hsieh, W.C., Wallach, D.A., Burrows, M., Chandra, T., Fikes, A., Gruber, R.E.: Bigtable: a distributed storage system for structured data. In: Proceedings of the 7th Symposium on Operating Systems Design and Implementation, pp. 205–218. OSDI 2006, USENIX Association, Berkeley, CA, USA (2006)
5. DeCandia, G., Hastorun, D., Jampani, M., Kakulapati, G., Lakshman, A., Pilchin, A., Sivasubramanian, S., Vosshall, P., Vogels, W.: Dynamo: amazon's highly available key-value store. SIGOPS Oper. Syst. Rev. **41**(6), 205–220 (2007)
6. George, L.: HBase: The Definitive Guide. O'Reilly Media Inc., Sebastopol (2011)
7. Holzschuher, F., Peinl, R.: Performance of graph query languages: comparison of cypher, gremlin and native access in neo4j. In: Proceedings of the Joint EDBT/ICDT 2013 Workshops, EDBT 2013, NY, USA, pp. 195–204. ACM, New York (2013)
8. Kostylev, E.V., Reutter, J.L., Vrgoc, D.: Containment of data graph queries. In: ICDT, pp. 131–142 (2014)
9. Kristina, C., Michael, D.: MongoDB: The Definitive Guide. O'Reilly Media, Sebastopol (2010)
10. Lakshman, A., Malik, P.: Cassandra: a decentralized structured storage system. ACM SIGOPS Oper. Syst. Rev. **44**(2), 35–40 (2010)
11. Neo4j.org: Neo4j - the world's leading graph database. http://www.neo4j.org/, Accessed on 16 june 2014
12. Nishimura, S., Das, S., Agrawal, D., Abbadi, A.: Md-hbase: a scalable multi-dimensional data infrastructure for location aware services. In: 2011 12th IEEE International Conference on Mobile Data Management (MDM), vol. 1, pp. 7–16, June 2011
13. Robinson, I., Webber, J., Eifrem, E.: Graph Databases. O'Reilly Media, Inc., Sebastopol (2013)
14. Shvachko, K., Kuang, H., Radia, S., Chansler, R.: The hadoop distributed file system. In: 2010 IEEE 26th Symposium on Mass Storage Systems and Technologies (MSST), pp. 1–10, May 2010
15. Thusoo, A., Sarma, J.S., Jain, N., Shao, Z., Chakka, P., Anthony, S., Liu, H., Wyckoff, P., Murthy, R.: Hive: a warehousing solution over a map-reduce framework. Proc. VLDB Endow. **2**(2), 1626–1629 (2009)
16. Vicknair, C., Macias, M., Zhao, Z., Nan, X., Chen, Y., Wilkins, D.: A comparison of a graph database and a relational database: a data provenance perspective. In: Proceedings of the 48th Annual Southeast Regional Conference, ACM SE 2010, NY, USA, pp. 42: 1–42: 6. ACM, New York (2010)
17. White, T.: Hadoop: The Definitive Guide, 3rd edn. O'Reilly Media Inc., Sebastopol (2012)
18. Wood, P.T.: Query languages for graph databases. SIGMOD Rec. **41**(1), 50–60 (2012)

Distributed XML Filtering Using HADOOP Framework

Panagiotis Antonellis, Christos Makris, and Georgios Pispirigos[✉]

Department of Computer Engineering and Informatics, Faculty of Engineering,
University of Patras, Patras, Greece
{adonel,makri,pispirig}@ceid.upatras.gr

Abstract. Publish-subscribe systems present the state of the art in information dissemination to multiple users. Current XML-based pub-sub systems provide users with considerable flexibility allowing the formulation of complex queries on the content as well as the structure of the streaming messages. Messages that contain one or more matches for a given user profile (query) are forwarded to the user. Typically the use of XML representation entails the profile representation with the use of the XPath query language and the employment of efficient heuristic techniques for constraining the complexity of the filtering mechanism. However, as the number of XML documents exchanged daily grows rapidly, the need for distributed management is becoming crucial. In this paper we propose three different approaches for distributed XML filtering using the Hadoop framework. The experimental results clearly demonstrate that the proposed techniques provide good scalability and effectiveness for very large number of document and user queries, compared to traditional XML filtering.

1 Introduction

The rapidly increasing volume of information (e.g., news feeds, data reports, advertisements) made available on the Internet has motivated the development of a new generation of applications based on selective data dissemination, where specific data is selectively deployed to a large number (e.g., millions) of distributed clients [1]. This trend has led to the emergence of novel middleware architectures that asynchronously distribute data from a set of publishers (i.e., data generators) to a large number of widely dispersed subscribers (i.e., data consumers) who have pre-registered their interest in specific information item, using a predefined set of filters/interests. In general, such publish-subscribe frameworks are implemented using a set of networked servers that selectively propagate relevant messages to the consumer population, where message relevance is determined by subscriptions representing the consumers' interests in specific messages.

Initial attempts to represent user filters/interests and match them against the incoming data items, typically employed "set of words" representations and keyword similarity techniques that are closely related to the well-known vector space model representation in the Information Retrieval area. These techniques, however, often suffer from limited ability to express user interests, being unable to fully capture the semantics of the user behavior and user interests. As an attempt to face this lack of richness in the

© Springer International Publishing Switzerland 2016
I. Karydis et al. (Eds.): ALGOCLOUD 2015, LNCS 9511, pp. 75–83, 2016.
DOI: 10.1007/978-3-319-29919-8_6

representation there have appeared lately [2, 3, 5, 6, 11] a number of systems that use XML representations for both documents and user profiles and that employ various filtering techniques to match the XML representations of user documents with the provided profiles.

The basic mechanism used to describe user profiles in XML format is through the XPath query language. XPath is a query language for addressing parts of an XML document, while also providing basic facilities for manipulation of strings, numbers and booleans. XPath models an XML document as a tree of nodes; there are different types of nodes, including element nodes, attribute nodes and text nodes and XPath defines a way to compute a string-value for each type of node.

2   Background

2.1   Related Work

In recent years, many approaches have been proposed for providing efficient filtering of XML data against large sets of user profiles. Depending on the way the user profiles and XML documents are represented and handled, the existing filtering systems can be categorized as follows:

Automata-Based Systems. Systems in this category utilize Finite State Automata (FSA) to quickly match the incoming XML document with the stored user profiles. While parsing the XML document, each node element causes one or more transitions in the underlying FSA, based on the element's name or tag. In XFilter [2], the user profiles are represented as queries using the XPath language and the filtering engine employs a sophisticated index structure and a modified Finite State Machine (FSM) approach to quickly locate and examine relevant profiles. A major drawback of XFilter is its lack of twig pattern support, as it handles only linear path expressions. Based on XFilter, a new system was proposed in [6] termed YFilter that combined all of the path queries into a single Nondeterministic Finite Automaton (NFA) and exploited commonality among user profiles by merging common prefixes of the user profile paths such that they were processed at most once. Unlike XFilter, YFilter handles twig patterns by decomposing them into separate linear paths and then performing post-processing over the intermediate matching results. In [16] a parallel implementation of YFilter for multi-core systems (shared-memory) is proposed by splitting the NFA into smaller parts, with each part assigned to a single thread. A distributed version of YFilter which also supports value-based predicates is presented in [13]. In this approach the NFA is distributed along the nodes of a DHT network to speed-up the filtering process and various pruning techniques are applied based on the defined value predicates on the stored user profiles. Finally in [9] a parallel XML filtering that supports fine-grained filtering of the incoming XML documents is described. A user can submit an hierarchy of filters and ever incoming XML document will be filtered against the stored filter hierarchies. In addition, the algorithm identifies exactly which parts of the incoming XML documents match with the filters of each user and only those parts are actually send to the user.

Sequence-Based Systems. Systems in this category encode both the user profiles and the XML documents as string sequences and then transform the problem of XML filtering into that of subsequence matching between the document and profile sequences. FiST [11] employs a novel holistic matching approach, that instead of splitting the twig patterns into separate linear paths, it transforms (through the use of the Prüfer sequence representation) the matching problem into a subsequence matching problem. In order to provide more efficient filtering, the user profiles sequences are indexed using hash structures. In XFIS [3] it is employed, a holistic matching approach which eliminates the need of extra post-processing of branch nodes by transforming the matching problem into a subsequence matching problem between the string sequence representation of user profiles and XML documents.

Stack-Based Systems. The representative system of this category is AFilter [5]. AFilter utilizes a stack structure while filtering the XML document against user profiles. Its novel filtering mechanism exploits both prefix and suffix commonalities across filter statements, avoids unnecessarily eager result/state enumerations (such as NFA enumerations of active states) and decouples memory management task from result enumeration to ensure correct results even when the memory is tight. XPush [10] translates the collection of filter statements into a single deterministic pushdown automaton using stacks. The XPush machine uses a SAX parser that simulates a bottom up computation and hence doesn't require the main memory representation of the document. XSQ [15] utilizes a hierarchical arrangement of pushdown transducers augmented with buffers. In [14], the author presents a system for evaluating XPath queries in a distributed environment, consisting of a large number of small mobile devices. Although the proposed system is efficient in such environments, it cannot be actually applied in a single multi-threading machine.

2.2   Paper Motivation and Contribution

In this paper we explore the problem on distributed XML filtering using the Hadoop framework. Especially, we propose three different distributed implementations of YFilter that utilize Hadoop to distribute the load to multiple nodes and thus greatly improve the efficiency of XML filtering. The experimental results clearly demonstrate the effectiveness of the proposed techniques with respect to the traditional non-distributed YFilter approach.

3   Hadoop Implementation of YFilter

In this section we describe three different techniques and implementations of YFilter using the Hadoop framework. Next, we perform a rich set of experiments in order to evaluate and compare the three approaches.

3.1   Implementation 1

The primary goal of the 1st implementation is to fully balance the processing workload that each processing slave, aka reducer is going to process. During the mapping process, the mapper splits the original XML file dataset, to equivalent subsets, regarding the number of XML files contained. The produced subsets are as many as the number of clusters' reducers. Each reducer processes its' assigned subset using the whole original XPATH queries profile, hence anyone going to construct and utilize the same NFA data structure. Despite the fact that this approach ensures a fair processing balance, the disadvantage of this implementation lies on the big NFA that each reducer should handle during processing, since the time needed per single XML is linearly increased as the constructed NFA grows.

3.2   Implementation 2

The 2nd approach aims to balance the processing workload by splitting the original set of XPATH queries profile. During the mapping process, the original constructed NFA, regarding its inherent structure, breaks into parts, the subNFAs. Each produced subNFA is assigned to a reducer. Each reducer, after merging all its assigned subNFAs to one, proceeds to filter the entire original XML input file set. Despite the fact that each reducer handles a much more manageable NFA, splitting the original XPATH queries set does not ensure a fair processing balance, since either the original NFA might not be able to break through equivalent parts due to its primary structure, or the different profile parts meets different processing requirements. Another disadvantage is that, each reducer should process the whole XML dataset, which is proven to be a very time consuming strategy.

3.3   Implementation 3

The 3rd implementation combines the advantages of both previous implementations. Specifically, during the mapping process, mapper not only splits the original constructed NFA structure to subNFAs but also defines the set of XML files which refer to and might be proven to interest, after complete processing, its respective produced subNFA. We use the same splitting technique as in 2nd implementation to define the produced subNFAs. In order to define each subNFA's set of possibly interesting XML files, mapper process the entire original XML dataset against an intermediate NFA structure, that is constructed from the merged common parts of all produced subNFAs, the common ancestor subNFA. The size of this intermediate structure depends on original NFA's native structure, but it is expected to be tremendously much more manageable than the original. Thus, the workload balance across reducers is not limited to the definition of XML files set that each one is going to process but also extended to the NFA structure that each should construct and manage during filtering process. This implementation theoretically seems to succeed the best workload balance, although inherits the disadvantage of the 2nd implementation where under certain circumstances, when either the root XML tag of a subNFA might be the "*" tag that matches any other tag or the split of subNFA revealed between nodes related with the ancestor-descendant type of relationship, a reducer might have to process the whole XML dataset but even though the size of its NFA would be much smaller than the used in a respective execution during the 2nd implementation.

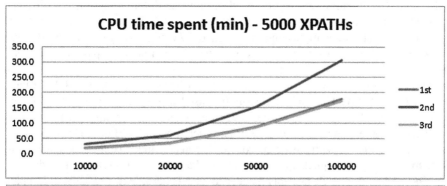

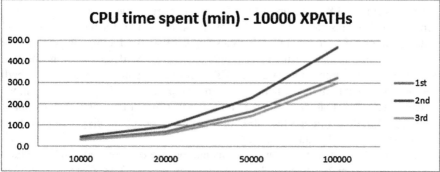

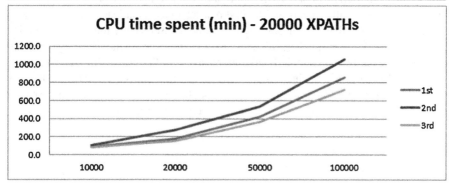

Fig. 1. CPU time spent (in minutes) for various Data sets of 10 K, 20 K, 50 K and 100 K of XML files, that all included files contained the requested tags.

3.4 Experiments

In our experiments, we use 2 different datasets generated by our own XML generator. The first one was 4.2 GB, consisted of 100.000 XML files which any of them could be possibly interesting, while the second one was 5.5 GB, consisted of 100.000 XML files that only half of them could be possibly interesting. Each generated XML file might contain 27 equiprobably selected XML tags and might have up to 100 different XML paths, where each path might be consisted of up to 25 different tags. The 4 different XPATH profile sets used were generated by our own XPATH profile generator. Each generated XPATH query might contain up to 25 of 27 equiprobably selected XML tags, where the possibility of existence an ancestor-descendant relationship between 2 XML tags was set to 5 %.

All the experiments were run on a cluster of 4 VirtualBox 4.3.28.r100309 Linux CentOS 6.6 VMs. W. The Hadoop distribution used was the 2.6.0, provided by Cloudera's CDH-5.4.0-1 parcel. All cluster's VMS were single processing thread machine at 1.2 GHz, where mapper allocated 8 GB and each reducer 2 GB of RAM (Fig. 1).

Comparing the 3rd with the 1st approach, we can conclude that the 3rd algorithm is in the average case 9 % quicker than the 1st one. Particularly, as far as the first dataset, where any XML file can be possibly meet each XPATH profiles criteria, the 3rd implementation is at least 3.8 % and up to 14.4 % times quicker. In the second dataset's case, where only half of inserted XML dataset might be interesting for the given XPATH profiles, is at least 6 % and up to 9.8 % times quicker than the 1st implementation. We can certainly come up to the result, that as either the inserted XML data set or the inserted size of the XPATH profile increases, the CPU processing gains increases, since both cases are equivalent to the increment of the required processing time. At this point, we should also underline that the gain rate in the 2 experiments differs, since the amount of interesting XML files, which is the subset of files that we obtain the bigger gains, differs in each respective XML dataset (Fig. 2).

Comparing the 3rd with the 2nd implementation, we can conclude with certainty from the presented graphs that the 3rd approach is in average 38.8 % times faster than the 2nd one. Specifically, regarding the first dataset's case is at least 32.9 % and up to 43.1 % times quicker than the 2nd, and in the second dataset's occasion is at least 48.5 % and up to 32.4 % times quicker. We can also observe that as the size of processing subNFA that each reducer should handle increases the CPU processing gains reduces. This is very logical, since the increment of the processing ssubNFA size is translated as increment of the required processing time. Although, this reduction factor can be easily reversed by extending the number of processing reducers (Fig. 3).

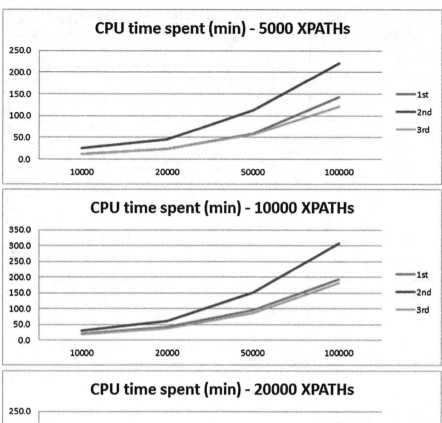

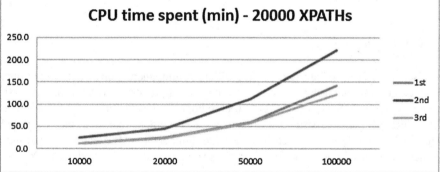

Fig. 2. CPU time spent (in minutes) for various Data sets of 10 K, 20 K, 50 K and 100 K of XML files, that half of included files contained the requested tags.

CPU processing Speedup for DataSet that all XMLs contains the requested tags						
Algorithm	# of XPATH	10000	20000	50000	100000	Avg
1st vs 2nd	5000	40.4%	40.3%	41.4%	41.3%	40.9%
	10000	26.4%	24.8%	29.3%	31.2%	27.9%
	20000	11.4%	35.4%	20.5%	19.4%	21.7%
2nd vs 3rd	5000	43.1%	42.2%	43.3%	43.9%	43.1%
	10000	34.6%	35.4%	38.0%	36.0%	36.0%
	20000	23.3%	44.3%	32.4%	31.8%	32.9%
3rd vs 1st	5000	4.6%	3.1%	3.2%	4.5%	3.8%
	10000	11.2%	14.0%	12.4%	7.0%	11.2%
	20000	13.4%	13.7%	14.9%	15.4%	14.4%

CPU processing % Speedup for DataSet that half of the XMLs contains the requested tags						
Algorithm	# of XPATH	10000	20000	50000	100000	Avg
1st vs 2nd	5000	51.9%	45.4%	46.8%	35.5%	44.9%
	10000	32.0%	30.2%	36.2%	37.4%	33.9%
	20000	43.0%	15.9%	19.5%	21.4%	25.0%
2nd vs 3rd	5000	53.1%	47.5%	48.6%	44.8%	48.5%
	10000	37.3%	37.9%	42.9%	41.2%	39.8%
	20000	47.2%	23.1%	27.1%	32.2%	32.4%
3rd vs 1st	5000	2.5%	3.9%	3.3%	14.4%	6.0%
	10000	7.8%	11.1%	10.5%	6.1%	8.9%
	20000	7.4%	8.5%	9.4%	13.7%	9.8%

Fig. 3. CPU processing time speedup for various Data sets of 10 K, 20 K, 50 K and 100 K of XML files, for all different implementations.

4   Conclusions

In this paper we have introduced three different implementations of the traditional YFilter algorithm that utilize the Hadoop framework for distributed XML filtering. All the approaches split the global NFA in order to improve the efficiency of the distributed XML filtering. The experimental results clearly demonstrate that the suggested approaches scale very well even in very large number of XML documents and user profiles, thus providing a boost in efficiency compared to the traditional YFilter approach.

References

1. Aguilera, M.K., Strom, R.E., Stunnan, D.C., Ashey, M., Chandra, T.D.: Matching events in a content-based subscription system. In: Proceedings of the ACM Symposium on Principles of Distributed Computing, PODC 1999, pp. 53–61 (1999)
2. Altinel, M., Franklin, M.L.J.: Efficient filtering of XML documents for selective dissemination of information. In: VLDB, pp. 53–64 (2000)
3. Antonellis, P., Makris, C.: XFIS: an XML filtering system based on string representation and matching. Int. J. Web Eng. Technol., IJWET 4(1), 70–94 (2008)

4. Budanitsky, A., Hirst, G.: Evaluating WordNet-based measures of lexical semantic relatedness. In: Association for Computational Linguistics, vol. 32, pp. 32–47 (2006)
5. Canadan, K., Hsiung, W., Chen, S., Tatemura, J., Agrrawal, D.: AFilter: adaptable XML filtering with prefix-caching and suffix-clustering. In: VLDB, pp. 559–570 (2006)
6. Diao, Y., Altinel, M., Franklin, M.L.J., Zhang, H., Fischer, P.: Path sharing and predicate evaluation for high-performance XML filtering. TODS **28**(4), 467–516 (2003)
7. Diaz, A.L., Lovell, D.: XML Generator. http://alphaworks.ibm.com/tech/xmlgenerator
8. Fellbaum, C. (ed.): WordNet, An Electronic Lexical Database. MIT Press, Cambridge (1998)
9. Grummt, E.: Fine-grained parallel XML filtering for content-based publish/subscribe systems. In: Proceedings of the 5th ACM International Conference on Distributed Event-Based System, DEBS 2011 (2011)
10. Gupta, A.K., Suciu, D.: Stream processing of XPath queries with predicates. In: SIGMOD, pp. 419–430 (2003)
11. Kwon, J., Rao, P., Moon, B., Lee, S.: FiST: scalable XML document filtering by sequencing twig patterns. In: VLDB, pp. 217–228 (2005)
12. Kwon, J., Rao, P., Moon, B., Lee, S.: Value-based predicate filtering of XML documents. Data Knowl. Eng. (KDE) **67**(1), 51–73 (2008)
13. Miliaraki, I., Koubarakis, M.: Distributed structural and value XML filtering. In: DEBS, pp. 2–13 (2010)
14. Olteanu, D.: SPEX: streamed and progressive evaluation of XPath. IEEE Trans. Knowl. Data Eng. **19**(7), 934–949 (2007)
15. Peng, F., Chawathe, S.: XSQ: a streaming XPath queries. TODS **30**, 577–623 (2005)
16. Zhang, Y., Pan, Y., Chiu, K.: A parallel XPath engine based on concurrent NFA execution. In: Proceedings of the IEEE 16th International Conference on Parallel and Distributed Systems, ICPADS 2010, pp. 314–321 (2010)

Efficient Bin Packing Algorithms for Resource Provisioning in the Cloud

Shahin Kamali[✉]

Massachusetts Institute of Technology, Cambridge, MA 02139, USA
skamali@mit.edu

Abstract. We consider the Infrastructure as a Service (IaaS) model for cloud service providers. This model can be abstracted as a form of online bin packing problem where bins represent physical machines and items represent virtual machines with dynamic load. The input to the problem is a sequence of operations each involving an insertion, deletion or updating the size of an item. The goal is to use live migration to achieve packings with a small number of active bins. Reducing the number of bins is critical for green computing and saving on energy costs. We introduce an algorithm, named HarmonicMix, that supports all operations and moves at most ten items per operation. The algorithm achieves a competitive ratio of 4/3, implying that the number of active bins at any stage of the algorithm is at most 4/3 times more than any offline algorithm that uses infinite migration. This is an improvement over a recent result of Song et al. [12] who introduced an algorithm, named VISBP, with a competitive ratio of 3/2. Our experiments indicate a considerable advantage for HarmonicMix over VISBP with respect to average-case performance. HarmonicMix is simple and runs as fast as classic bin packing algorithms such as Best Fit and First Fit; this makes the algorithm suitable for practical purposes.

1 Introduction

We consider Infrastructure as a Service (IaaS) model in the cloud which has received increasing attention in the past few years. In this model, a cloud service provider such as Amazon EC2 rents virtual machines (VMs) to clients. Each VM is capable of running several applications with dynamic loads that vary by the time. The total load of applications encapsulated in a VM defines the *load* of the VM. The applications are unpredictable in the sense that their load and the pattern of their changes cannot be predicted in advance. In other words, the load of VMs is not known beforehand and changes over time. A service provider has to assign VMs into physical machines (PM's) so that the total load of all VMs in each machine is no more than the uniform capacity of PM's. In other words, servers should not be overloaded in order to avoid bottlenecks in the system and to balance the load between PMs. Moreover, the number of PM's that are used to host VMs is desired to be as small as possible. This objective is important for green computing and reducing energy costs. Particularly, inactive PMs which do

© Springer International Publishing Switzerland 2016
I. Karydis et al. (Eds.): ALGOCLOUD 2015, LNCS 9511, pp. 84–98, 2016.
DOI: 10.1007/978-3-319-29919-8_7

not host any VM can hibernate in fractions of a second [10] and hence save on energy costs. The IaaS model, as described above, is closely related to the classic bin packing problem.

In the bin packing problem, the input is a set of *items* each having a *size* in the range (0,1]. The goal is to place these items into a minimum number of bins of uniform capacity. In the IaaS model, items represent VMs; item sizes represent the load of VMs; and bins represents PMs. The bin packing problem requires the total size of items in each bin to be at most equal to the unique capacity of bins, and the objective is to place items into a minimum number of bins. In the online version of the problem, item sizes are not known in advance. Instead, they form a sequence that is revealed item by item. An online algorithm should place an item into a bin without any knowledge about the forthcoming items. In the IaaS model, VMs' loads are not known in advance; hence, the online bin packing is more relevant compared to the offline bin packing where item sizes are known beforehand. An example of an online bin packing algorithm is Next Fit which keeps one *active* bin and places an incoming item in the active bin if it has enough space; otherwise, it closes the bin and opens a new active bin. First Fit is another online algorithm that places each item into the first bin, in the order that they are opened, which has enough space (and opens a new bin if required). Best Fit is similar to First Fit except that it maintains bins in the decreasing order of their *levels*, where the level of a bin is total size of items in it. Harmonic algorithm has a parameter K, where K is a positive integer, and partitions the unique interval into sub-intervals $(1/2, 1], (1/3, 1/2], \ldots, (1/(K + 1), 1/K], (0, 1/K]$, and applies a separate Next Fit strategy for items with sizes in each sub-interval.

Competitive analysis is the standard approach for comparing online algorithms. For an online algorithm $\mathbb{A}$, we use $\mathbb{A}(\sigma)$ to denote the number of bins opened by $\mathbb{A}$ for packing a sequence σ. Similarly, we use $\text{OPT}(\sigma)$ to denote the number of bins opened by an optimal algorithm OPT for packing σ. In the asymptotic sense, the value of $\text{OPT}(\sigma)$ is assumed to be large and the *asymptotic* competitive ratio of $\mathbb{A}$ is defined as the maximum value of $\frac{\mathbb{A}(\sigma)}{\text{OPT}(\sigma)}$ for any sequence σ[1] . Next Fit has a competitive ratio of 2, Best Fit and First Fit both have competitive ratio 1.7 [7], and the competitive ratio of Harmonic converges to 1.69 for large values of K [8].

In the standard setting for online bin packing, the decisions of an online algorithm are irrevocable and an item in a bin B cannot be moved to another bin B'. In the IaaS model, however, the dynamic size of VMs requires moving them between servers to avoid overloaded PMs (e.g., when the load all VMs hosted by a PM increase). *Live migration* [3] enables moving VMs between PMs without interrupting applications running inside them. Different strategies are introduced for live migration (see, for example, Sandpiper [13] and VectorDot [11]). However, these approaches are merely focused on load balancing and do

---

[1] Throughout the paper, by competitive ratio, we mean asymptotic competitive ratio. For results related to the *absolute* competitive ratio of bin packing algorithms, we refer the reader to [4,5].

not consider green computing. In this paper, we study the online bin packing algorithms for the IaaS model, which is defined as follows.

Definition 1. *In the IaaS model of bin packing, the input is an online sequence of operations on items (VMs) with sizes (loads) in the range $(0, 1]$. Each operation involves either inserting an item to any bin (PM), removing an item from a given bin, or updating the size of an item from x to y. Upon applying each operation, an online algorithm can use live migration to move a constant number of items between bins. The size of items in each bin at any given time should not be more than the uniform capacity of bins. The goal is to achieve packings with minimum number of bins (active PMs).*

1.1 Previous Work and Contribution

Gambosi et al. [6] studied a version of online bin packing where only insertion and deletion are allowed. They introduced an online algorithm with a competitive ratio of at most 4/3. Unfortunately, their algorithm is quite complicated and does not seem suitable for practical purposes. Moreover, it does not support update operation. An update operation can be simulated with a delete and then an insert operation. However, as pointed out in [12], this might require moving a large number of items between bins. For example, consider a packing in which there are two bins each having an item of size $0.5 + \epsilon$ and $0.5/\epsilon - 1$ items of size ϵ. Other bins in the packing each include $1/\epsilon - 1$ items of size ϵ. If the size of one of the items of size $0.5 + \epsilon$ increases to $0.5 + 2\epsilon$, its bin gets overloaded. To fix the packing, it suffices to move an item of size ϵ to another bin. However, if we delete and re-insert the updated item, either an extra bins should be opened or at least $0.5/\epsilon$ items should be moved.

The IaaS model of bin packing has been recently studied by Songe et al. [12]. There, the authors introduced an algorithm, called bin packing with variable-sized items (VISBP), which has a competitive ratio of 1.5 and supports all three operation. Although this algorithm uses live migration to improve over the lower bound 1.54037 [1] for competitive ratio of purely online algorithms, it leaves a lot of space for improvement. In particular, we show that live migration can be used more effectively to improve over the competitive ratio and, more importantly, the average-case performance of VISBP.

In this paper, we apply more complicated packing techniques to introduce an algorithm, called HarmonicMix, for the IaaS model. Recall that Harmonic algorithm for bin packing includes i item in the range $(1/(i+1), 1/i]$ in a bin of type i. This particular structure makes the algorithm suitable for dynamic packing as items of same type can replace each other in their harmonic bin. Unfortunately, Harmonic algorithm has a poor average-case performance [9] when compared to classic algorithms such as Best Fit and First Fit. To address this issue, in HarmonicMix, we make use of live migration to improve the packings of Harmonic algorithms. This ensures a good average-case performance in terms of the number of active bins. At the same time, the algorithm moves a small number of items, at most ten items, per operation. We prove that the competitive ratio of HarmonicMix is $4/3$, which is better than $3/2$ of VISBP. To compare average-case performance of these algorithms, similar to many related works for average-case study of bin packing algorithms (see [4] for a review), we test the algorithms on randomly-generated input sequences. Our experiments indicate that HarmonicMix has an advantage over VISBP, not only in the worst-case, but also in the average case.

2   HarmonicMix Algorithm

In this section, we introduce and analyse the HarmonicMix algorithm. Similar to the previous works on dynamic bin packing (see, e.g., [6,12]), we assume item sizes are larger than a fixed value. For example, in the VISBP algorithm of [12], this fixed value is defined to be $1/6$. Items with sizes at most $1/6$ are grouped together to form *multi-items* with sizes in the range $(1/6,1/3]$. For HarmonicMix, we define this fixed value to be $1/8$, i.e., we group items of size at most $1/8$ into multi-items with sizes in the range $(1/8,1/4]$. This can be done with no computing overhead. In what follows, we always assume item sizes are larger than $1/8$. Before introducing the algorithm, we describe a general family of packings called *valid packings*. We prove that any algorithm that maintain a valid packing has a competitive ratio of at most $4/3$. Later, we will show how to maintain a valid packing by moving a small number of items after each operation.

2.1   Valid Packings

In what follows, we refer to an item as being *large* if it is larger than $1/2$, *medium* if it is in the range $(1/3,1/2]$, *small* if it is in the range $(1/4,1/3]$, and *tiny* if it is in the range $(1/8,1/4]$. We correspond each bin with the largest item in the bin. For example, a bin is medium if it includes a medium item and no large item. A given packing of n items is valid if the following conditions hold. We use the term 'almost all' for a set of bins to indicate all bins in the set except potentially a constant number of them.

1. Almost all medium bins include two medium items and possibly one tiny item.
2. Almost all small bins include three small items.
3. Almost all tiny bins have a level of at least $3/4$.
4. For almost all large bins like B that does not contain a medium or small item, and for any medium or small item y in bins other than large bins, we have $x + y > 1$, where x is the size of the large item in B.
5. For almost all large or medium bins like B either $level(B) \geq 3/4$ or there is no tiny bin in the packing.

Intuitively, properties 1–3 can be maintained by placing medium, large, and tiny items in separate bins in a similar fashion as the Harmonic algorithm does. Property 4 implies that a large item should be accompanied with a medium or a small if possible; if it is not possible, then the item might be accompanied by tiny items. Property 5 implies that tiny items should be placed in large and medium bins to ensure a level of at least 3/4 for these bins. In other words, there is a tiny bin in the packing only when the level of all large and medium bins is 3/4 or more.

Lemma 1. *Any algorithm $\mathbb{A}$ that maintains a valid packing has a competitive ratio of at most 4/3.*

Proof. We consider the following two cases for a valid packing P of an input sequence σ and prove the claim for each case separately.

Case I: Assume P includes a tiny bin. We prove that the level of all bins, except possibly a constant number of them, is at least 3/4. This gives a competitive ratio of 4/3 for the packing. Property 2 implies that almost all small bins include three small items, i.e., they have a level in the range (3/4,1]. Property 3 indicates that tiny bins also have level 3/4 or more. Property 5 implies that any large or medium bin B has level 3/4 or more; otherwise, any of the tiny items placed in the tiny bin should have been placed in B.

Case II: Assume P does not include a tiny bin. Consider a fixed optimal packing of σ. In this packing, we refer to the large items that are accompanied by a small or a medium item as *blue* large items and refer to the rest of large items as *red* large items. For each item x, we define a *weight* $w(x)$ for x as follows. For a red large $w(x) = 1$, for a blue large item $w(x) = 5/6$, for a medium item $w(x) = 1/2$, for a small item $w(x) = 1/3$, and for tiny items $w(x) = 0$. Let $W(\sigma)$ denote the total weight of items in σ. We prove $\mathbb{A}(\sigma) \leq W(\sigma) + c$ for some constant c and $\mathrm{OPT}(\sigma) \geq 3/4 \times W(\sigma)$. From these two inequalities, we conclude $\mathbb{A}(\sigma) \leq 4/3 \times W(\sigma) + c$, which completes the proof.

First, we prove $\mathbb{A}(\sigma) \leq W(\sigma) + c$. We show that items in almost all bins in the packing of $\mathbb{A}$ have an average total weight of at least 1. By property 1, almost all medium bins include two medium items. The total weight of these items would be $2 \times 1/2 = 1$. Similarly, by property 3, almost all small bins include three items, each with weight 1/3. The total weight would be $3 \times 1/3 = 1$. There is no tiny bin in Case 2. It remains to show average the weight of items in large bins is at least 1. Let R and B respectively denote the number of red and blue large items; the total contribution of large items to the total weight of all large bins is $R + 5/6 \cdot B$. We claim that at least $B/2$ of large bins include a medium of a small items. If that is true, the contribution of these small/medium items to the weight of large bins would be at least $B/2 \times 1/3 = B/6$. Hence, the total weight of items in large bins would be at least $R + 5/6 \cdot B + B/6 = R + B$. Since the algorithm opens $R + B$ large bins, the average weight of large bins would be at least 1. To prove the claim, we consider set S_l formed by large items and set S_{ms} formed by the union of medium and small items in the input sequence. By definition of blue bins, we have $|S_l| \geq B$ and $|S_{ms}| \geq B$. Let S_l^* and S_{ms}^* respectively denote the smallest $\lfloor B/2 \rfloor$ items of S_l and S_{ms}. For any pair $(x, y), x \in S_l^*, y \in S_{ms}^*$, we have $x + y \leq 1$. To sea that, consider the blue large item z which has median size among the blue large items. There are roughly $B/2$ blue large items smaller than z and $B/2$ small/medium items smaller than $1 - z$. So, all $B/2$ items of S_l^* fit with all $B/2$ items of S_{ms}^*. By property 4, the algorithm tends to place medium/small items in large bins (and for that, they have priority over tiny items). Hence, at least $B/2$ medium/small items are placed in large bins and the claim follows.

Next, we prove $\text{OPT}(\sigma) \geq 3/4 W(\sigma)$. We show that any given bin in the fixed optimal packing has a total weight of at most $4/3$. By definition, bins with red large items in the optimal packing do not include medium or small items. They might contain tiny items which do not contribute to the total weight. So, the total weight of large bins with a red item is one. Next, consider bins in the optimal packing that includes a blue large item. These bins include at most one other item, i.e., a medium or a small item (tiny items are ignored as their weight is zero). In the former case, the weight of the bin would be $5/6 + 1/2 = 4/3$. In the latter case, the weight would be $5/6 + 1/3 < 4/3$. Next, assume a bin without large items. It might contain 1) two medium items with total weight of one; 2) one medium and two small items with total weight $1/2 + 2/3 < 4/3$; 3) three small items with a total weight of one. In all cases, the total weight of bins is at most $4/3$. So, the total weight W of all items is distributed between at last $3/4 \cdot W$ bins.   □

2.2   Nice Packings

The HarmonicMix algorithm maintains a certain type of valid packings, called *nice packings*, which we describe here. By property 1 of valid packings, almost all medium bins include two medium items. These two items would have a total size in the range $(2/3,1]$. In order to fulfill property 5 of a valid packing, the bin might also include a tiny bin. This implies that each medium bin has two *spots* for two medium items and one spot for a tiny item. The tiny spot might be empty but each medium spot includes a medium item. Similarly, property 2 indicates that small bins include three spots for small items, and in almost all small bins the three spots are occupied (i.e., there is no empty spot). Each large bin includes a non-empty spot for a large item. There is also a spot for a medium or a small item. We call this spot medium/small spot (there is no priority between medium and small items for occupying this spot). There are also two tiny spots which are filled with tiny items only if the medium/small spot is empty. This implies that if there is a medium/small item that can fill the medium/small spot, the tiny spots need to be empty.

Note that the above description for bin spots is consistent with the definition of a valid packing. In a nice packing, in addition to the five properties of valid packings, we require that a tiny item be placed in the tiny spot of a large or medium bin, and only if it is not possible then it is placed in a tiny bin. As an example, consider a large bin B which includes an item of size $5/6 - \epsilon$ for some small positive value of ϵ. Clearly, no medium/small item fits in the remaining space of the bin. However, a tiny item x of size $1/6 + \epsilon$ does fit in B. For a valid packing, it is not required to place x in B because property five holds (since B has level more than $3/4$). However, to have a nice packing, we require x to be

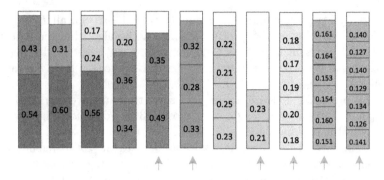

Fig. 1. An example of a nice packing. Colors of items indicate their types (red for large, blue for medium, amber for small, and different shades of green for subfamilies of tiny items). The arrows point to active bins of different groups. Assume we remove the medium item 0.43 from the first bin. To maintain a nice packing, the empty medium/small spot should be filled with a medium item (e.g., 0.35 in the active medium bin) or a small item (e.g., 0.32 in the active small bin) (Color figure online).

in the tiny spot of B rather than a tiny bin. This property of nice packings is not used in our worst-case analysis; however, it is essential for having a good average-case performance.

In addition to the above-mentioned property, in a nice packing, tiny bins are more structured in the following sense. Recall that an item is tiny if it is in the range $(1/8, 1/4)$. We further partition this interval into sub-intervals $T_1 = (1/5, 1/4]$, $T_2 = (1/6, 1/5]$, $T_3 = (1/7, 1/6]$, and $T_4 = (1/8, 1/7]$. Each tiny bin of a nice packing includes tiny items of the same intervals. We say a tiny bin has type T_i if it includes tiny items of type T_i ($i \in \{1, 2, 3, 4\}$). At any given time, all tiny bins of type T_i include $i + 3$ items of type T_i. The only exception is the most recently opened bin, called the *active bin of type T_i*, which might include less than $i + 3$ items. This way, all tiny bins, except four of them (the active bins), have level $4/5$ or more, which ensures property 3 holds. We extend the notion of active bins to medium and small items. Recall that there are two medium spots in almost all medium bins. The only potential exception is the most recently opened bin, which we call the *active medium bin*. Similarly, there are three small spots in each small bin, again, with the exception of one *active small bin*. Figure 1 provides an illustration of a nice packings.

Any of the insertion/deletion/update operations might result in a packing which is not nice any more. To fix this, we apply live migration to maintain a nice packing. In many cases, this involves moving an item from an active bin to another bin of the same type. This might result in an empty active bin; in this case we declare another bin of the same type as the new active bin. Similarly, upon inserting an item to the active bin, we might need to open a new bin and declare it as the new active bin.

2.3   Insert/Delete Operations

In what follows, we describe how HarmonicMix updates maintains a nice packing after an insert or a delete operation.

Lemma 2. *It is possible to maintain a nice packing after an insert/delete operation by moving at most five items per operation.*

Proof. To prove the lemma, we discuss how each operation separately.

insert-tiny (no move). Assume we want to insert a new tiny item of size x. To maintain a nice packing, first we check if there is a tiny spot in large or medium bins in which x fits. If there is, we place x there; otherwise, we place x into the active bin of its type. No item is moved from the packing. If x does not fit into the active bin, the level of the bin is more than $3/4$ and property 3 holds. One can easily check that the other properties of a valid packing also hold. Note that no item is moved as a result of this operation.

remove-tiny (two moves). Assume we want to remove a tiny item x from a bin B. If B is an active bin, we remove x and no other item is moved. The packing remains valid since the properties of valid packing do not apply to active bins. If B is a non-active tiny bin, we fill the empty spot of x in B with an item x' in the active bin of the same type as B. Since x and x' belong to the same sub-class of tiny items, the packing remains valid. Next, assume x is in the tiny spot of a large or a medium bin. If we can replace x with another item x' located in a tiny bin, we move x's to B. This might require moving another item x'', from the active bin of the same sub-class of x', to the empty spot of x'. In total, at most two items are moved.

insert-small (two moves). To insert a small item x, we first check if there is a large bin B with an empty medium/small spot in which x fits. If there is, we place x in B and remove at most two items from the tiny spots of B and re-insert them; this would require moving at most two items. If there is no such large bin B in which x fits, we simply place x in the active small bin.

remove-small (five moves). If we remove a small item from the active small bin, the packing remains nice and no item is moved. To remove a small item x from a non-active small bin B, we simply replace x with a small item x' in the active small bin. This requires moving one item, i.e., x'.

Next, assume a small item x is removed from a large bin B. We might need to fill the empty spot of B with a medium or a small item in a non-large bin. If a small item is moved, as discussed above, at most one other item is moved, i.e., two items are moved in total. If a medium item is moved, as will be discussed later (see the first paragraph of remove-medium), we might need to move at most four other items to fix the packing, i.e., at most five items are moved in total.

insert-medium (two moves). To insert a medium item x, we first check to see if there is an empty medium/small spots in any of the large bins in which x fits. If there is such spot in a bin B, we place x in B. In case B includes one or two tiny items, we remove them from B and re-insert them to the packing. As suggested above, no item is moved after inserting tiny items. So, at most two items are moved. Next, assume there is no spot for x in the large bins. We place x in the active medium bin and open a new bin if required. In case we open a new bin, the previous active bin will have an empty spot which might be filled with an item in a tiny bin. Removing an reinserting such tiny item requires at most two moves.

remove-medium (five moves). Assume a medium item x is removed from a medium bin B. If B is the active medium bin, no item needs to be moved. If B is a non-active bin, the empty spot of x in B is filled with a medium item x' from the active bin. This might result in an overloaded bin when $x' > x$; this happens only if there is a tiny item z in B. To fix the packing, we remove z from B and re-insert it; this only requires moving z since no item is moved after re-inserting a tiny item. To ensure a nice packing, we might need to fill the empty spot of z with another tiny item z' in a tiny bin B'. This requires filling the empty spot of z' in B' with another tiny item z'' in the active tiny bin of the same sub-class. In total, we moved at most four items (x', z, z', and z'').

Next, assume a medium item x is removed from a large bin B. To maintain a valid packing, we might need to fill the empty spot of B with a medium or a small item from the non-large bins. If a medium item is moved, as discussed in the above paragraph, we might need to move at most four other items to fix the packing, i.e., at most five items in total. If a small item is moved, at most one other item is moved, i.e., two items are moved in total.

insert-large (five moves). Assume we want to insert a large item x. We open a new bin B for x; this bin would have an empty medium/small spot. If there is a medium/small item y in non-large bins which fits in the empty spot, we move y to B. As discussed earlier, removing a medium or small item from a non-large bin requires moving at most four other items, i.e., we move at most five items in total.

If there is no medium or small item that fits in the medium/small spot, we move tiny items into the two tiny spots of B. This requires moving at most two tiny items from tiny bins, and each potentially need moving another tiny item from the active bin of the same sub-class to the the new empty spots. In total, we move at most four items.

remove-large (three moves). Assume we remove a large item x from a large bin B. To maintain a nice packing, we remove other items in B and re-insert them to the packing. Assume the medium/small spot in B is occupied with an item y. Re-inserting y to the packing requires moving at most two other items. In total, at most three items are moved. If the medium/small spot in B is empty, there are at most two items in the tiny spots of B. Removing and re-inserting these items requires at most two moves.   □

2.4   Update Operations

A simple approach to implement updates is to remove the updated item and re-insert it. By Lemma 2, each operation requires moving at most five items. So, this approach requires moving at most ten items per update. While in the worst-case we might indeed need ten moves, for most cases, we can maintain a nice packing with less overhead.

Lemma 3. *Updating the size of an item so that it becomes or remains a tiny item requires at most five moves. If an item becomes or remains small or medium, at most seven moves are required. If a tiny items becomes large, at most seven moves are required. If a medium or small item becomes large, at most ten moves are required. If a large item remains large after an update, at most six moves are required.*

Proof. As before, we discuss each operation separately.

update-to-tiny (five moves). Assume we update the size of an item x so that it becomes tiny after the update. We remove x from the packing (at most five moves) and re-insert it (no move). In total, we move at most five items.

update-to-small and update-to-medium (seven moves). Assume we update the size of an item x so that it becomes small or medium after the update. We remove x from the packing (at most five moves) and re-insert it (at most two moves). In total, we move at most seven items.

update-to-large (ten moves). Assume we update the size of an item x so that it becomes large after the update. If x is tiny before the update, we remove x, using at most two moves, and re-insert it, using at most five moves. The number of moves will be at most seven.

Assume x was a small item in a small bin B before the update. After the update, we remove x (the first move), and fill its empty spot with an item from the active small bin (the second move), and then re-insert x as a large item (at most five additional moves). In total, at most seven moves are required. If x was a small item in a large bin before the update, we remove and re-insert it with at most ten moves.

Assume x was a medium item in a medium bin B before the update. So, x was placed with another medium item x' in B. If we have $x + x' \leq 1$ after the update, we maintain the nice packing by removing and re-inserting the tiny item of B (if there is one). That would require moving one item. If $x + x' > 1$ after the update, we need to remove x' from B. If there is another medium item x'' in a medium bin B'' so that $x + x'' \leq 1$, we swap x and x' (the first two moves). Moreover, if there was an item in the tiny spot of B, we remove and re-insert that item (the third move). Finally, if B'' is overloaded after the insertion, we remove and re-insert the tiny items in it (the fourth move). In that case, the tiny spot of B'' might be filled with another tiny item in a tiny bin (the fifth move). The resulting empty tiny spot in the tiny bin might be filled with an item in the active bin of the same sub-class (the sixth move).

Assume x was a medium item in a large bin B before the update. We simply remove x from B (at most five moves) and re-insert it as a large item (at most five moves). In fact, there are instances in which, to maintain a nice packing, these ten moves are required to maintain a nice packing.

Assume x was large before the update and its size is increased. If the medium/small spot is empty, we might need to move and replace at most two tiny items from the bin using at most six moves. Next, assume the medium/small spot is filled with an item y. After the increase, if the bin B is not overloaded, the packing is still nice. Otherwise, we need to remove y from B. Assume there is another medium item y', with size smaller than y, that can replace y in B. Swapping y and y' requires two moves. After that, the bin B' of y' might become overloaded. In that case, we remove the tiny item in B', denoted by z', and re-insert it to the packing (the third move). To fill the empty spot of z' in B', we might need to move two more tiny items (the fourth and the fifth moves). In total, we move at most five items. If there is no item y' to swap with y, we remove and re-insert y using two moves. Moreover, at most two tiny items will be moved to B, each requiring moving at most one other item from the active bin of their sub-class. In total, we move at most six items.

Next, assume x was large before the update and its size is decreased. If the medium/small spot was occupied before the update, the packing remains nice (no move). Otherwise, we should check if there is a medium or small item y in non-large bins that can fill the medium/small spot. If there is, we move y to B (the first move). The empty spot of y in its previous bin B' will be replaced by another item y' from the active medium/small bin (the second move). If y is medium, this might cause an overflow in B' which can be fixed by moving the tiny item z of B to the active bin of the same type (the third move) and replacing it with another tiny item z' (the fourth move). There might be tiny items in the tiny spots of B before the update. We have to remove these items from B and re-insert them to the packing (the fifth and the sixth moves). Note that inserting a tiny item does not cause extra moves. If there is no medium/small item y to be placed in B after the update, the tiny spots in B are filled with items in tiny bins, using at most four moves. In total, we move at most six items when the size of a large item decreases.   □

We use the case analysis presented in Lemmas 2 and 3 to devise the HarmonicMix algorithm. From Lemmas 1, 2 and 3, we conclude the following theorem:

Theorem 1. *HarmonicMix has a competitive ratio of at most 4/3 for the IaaS model of bin packing. The number of items moved for each operation is at most ten.*

3   Experiments

The results in the previous section imply that, in the worst case, HarmonicMix has an advantage over VISBP, i.e., it opens a smaller number of bins. In this section, we experimentally compare the two algorithms to study their average-case performance. We generate random sequence of operations which involve items whose sizes are also generated independently at random. The size of items is in defined to be in the range (1/6,1] so that HarmonicMix and VISBP work on the same set of items (otherwise, multi-items of the two algorithms will be different).

We present three experiments. In the first experiment, item sizes are generated uniformly at random from the range (1/6,1]. In the second experiment, item sizes follow a normal distribution with mean 0.5 and standard deviation 1. In the third experiments, item sizes take values from the set $\{1/1000, 2/1000, \ldots, 1\}$ following a Zipfian distribution with skew parameter $s = 1.1$ (the smaller the size, the more frequent the item). Each experiment has five phases. Each phase includes $n = 50,000$ operations. In phase one, n random numbers are inserted to the packing. No item is removed or updated during this *insert-only phase*. In the second phase, with a chance of 90 percent an insert operation is performed while chances of delete and update are equal to 5 percent. We call this the *insert-intensive phase*. In the third phase, with a chance of 50 percent, an insert operation is applied and with a chance of 50 percent, a delete or an update operation is applied (each with the same chance of 25 percent). We call this phase, *equal-load phase*. In the fourth phase, called *delete/update intensive phase*, with a chance of 10 percent an insert is applied and with a chance of 90 percent, a delete or update operation is applied (each with the same chance of 45 percent). Finally, in the last phase, called *delete-only phase*, only delete operations are applied.

Next, we describe how these insert/delete/update operations are defined. On an insertion, a new item of random size is inserted to the packing. On a deletion, an existing item in the packing is selected uniformly at random and removed. On an update, an existing item is selected randomly and a random value in the range $(-0.1, +0.1)$ is added to the size of x. If the size of x becomes larger than 1 or smaller than 1/6, another random change is added to have the size in the desired range.

Besides VISBP and HarmonicMix, we also modified Best Fit and First Fit to include them in our experiments. In contrast to the classic algorithms, our modified algorithms use live migration when the size of an item is updated (increased) so that the bin is overloaded, i.e., its level is more than 1. Only in this case, the item with the updated size is removed and re-inserted in the packing. Moreover, when the level of a bin becomes zero (when all items in the bin are removed) the bin is removed from the packing, i.e., it is not counted as one of the bins used by the algorithms.

Figure 2 show the number of bins for each algorithm at the end of each phase. One interesting observation is that the packings of Best Fit and First Fit are better than VISBP at the initial phases. This is partially because VISBP is designed in a way to improve the worst-case performance with a minimum number of live migrations. Moreover, Best Fit and First Fit algorithm are known to be optimal algorithms for packing sequences that are generated randomly (see, e.g., [2]). In particular, when items are removed and new random items, the resulting empty spots in the packings of Best Fit and First Fit is likely to be filled with new items (which have the same distribution on their sizes). In the last two phases in which delete operations are more frequent than inserts, VISBP

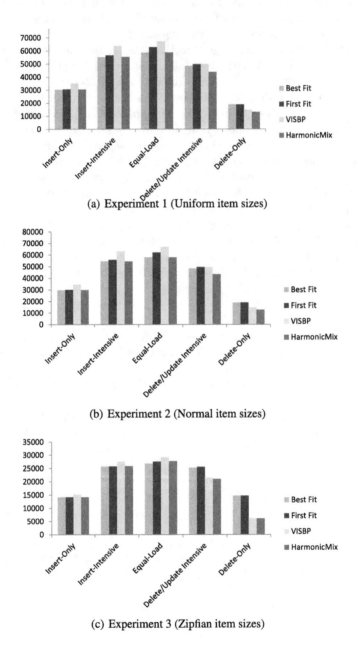

(a) Experiment 1 (Uniform item sizes)

(b) Experiment 2 (Normal item sizes)

(c) Experiment 3 (Zipfian item sizes)

Fig. 2. The average number of active bins at the end of each phase of the experiments.

shows its advantage over Best Fit and First Fit. HarmonicMix has advantage over other algorithms in almost all phases. The algorithm has a visible advantage over VISBP, and its advantage over Best Fit and First Fit is evident in the last two phases.

We also counted the total number of times that items have been moved in VISBP and HarmonicMix. The average number of migrations per operation in experiment 1 is 0.234 for VISBP and 0.457 for HarmonicMix. Similar numbers are observed for experiments 2 and 3. One can conclude that HarmonicMix tends to move more items to improve the quality of its packing while VISBP tends to minimize the number of migrations instead. Note that, although VISBP/HarmonicMix move at most seven/ten items per operation in the worst case, the expected number of moves is much smaller.

4   Concluding Remarks

HarmonicMix maintains valid packings that are also nice packings. The algorithm can be modified to maintain valid packings, which are not nice, while moving at most five items per operation. Such algorithm has the same competitive ratio of 4/3 of HarmonicMix. So, in the worst case, the algorithm has an advantage over VISBP and HarmonicMix. However, the new algorithm performs poorly on average. We leave further analysis of this algorithm and other variants of HarmonicMix as a future work.

References

1. Balogh, J., Békési, J., Galambos, G.: New lower bounds for certain classes of bin packing algorithms. Theor. Comput. Sci. **440–441**, 1–13 (2012)
2. Bentley, J.L., Johnson, D.S., Leighton, F.T., McGeoch, C.C., McGeoch, L.A.: Some unexpected expected behavior results for bin packing. In: Proceedings of 16th Symposium on Theory of Computing (STOC), pp. 279–288 (1984)
3. Clark, C., Fraser, K., Hand, S., Hansen, J.G., Jul, E., Limpach, C., Pratt, I., Warfield, A.: Live migration of virtual machines. In: 2nd Symposium on Networked Systems Design and Implementation (NSDI) (2005)
4. Coffman, E.G., Garey, M.R., Johnson, D.S.: Approximation algorithms for bin packing: a survey. In: Approximation Algorithms for NP-hard Problems. PWS Publishing Co (1997)
5. Coffman Jr., E.G., Csirik, J., Galambos, G., Martello, S., Vigo, D.: Bin packing approximation algorithms: survey and classification. In: Pardalos, P.M., Du, D.Z., Graham, R.L. (eds.) Handbook of Combinatorial Optimization, pp. 455–531. Springer, New York (2013)
6. Gambosi, G., Postiglione, A., Talamo, M.: Algorithms for the relaxed online bin-packing model. SIAM J. Comput. **30**(5), 1532–1551 (2000)
7. Johnson, D.S.: Near-optimal bin packing algorithms. Ph.D. thesis, MIT (1973)
8. Lee, C.C., Lee, D.T.: A simple online bin packing algorithm. J. ACM **32**, 562–572 (1985)

9. Lee, C.C., Lee, D.T.: Robust online bin packing algorithms. Technical report 83–03-FC-02, Department of Electrical Engineering and Computer Science, Northwestern University (1987)
10. Meisner, D., Gold, B.T., Wenisch, T.F.: Powernap: eliminating server idle power. In: Proceedings of the 14th International Conference on Architectural Support for Programming Languages and Operating Systems, pp. 205–216 (2009)
11. Singh, A., Korupolu, M.R., Mohapatra, D.: Server-storage virtualization: integration and load balancing in data centers. In: Proceedings of the ACM/IEEE Conference on High Performance Computing, pp. 53 (2008)
12. Song, W., Xiao, Z., Chen, Q., Luo, H.: Adaptive resource provisioning for the cloud using online bin packing. IEEE Trans. Comput. **63**(11), 2647–2660 (2014)
13. Wood, T., Shenoy, P.J., Venkataramani, A., Yousif, M.S.: Sandpiper: black-box and gray-box resource management for virtual machines. Comput. Networks **53**(17), 2923–2938 (2009)

Transaction Management for Cloud-Based Graph Databases

Georgia Koloniari[1(✉)] and Evaggelia Pitoura[2]

[1] Applied Informatics Department, University of Macedonia, Thessaloniki, Greece
gkoloniari@uom.gr
[2] Department of Computer Science and Engineering,
University of Ioannina, Ioannina, Greece
pitoura@cs.uoi.gr

Abstract. Many graph databases, both open and proprietary, have been recently developed to efficiently store and manage graph structured data. As the volume of such data grows, graph databases most often offer distributed solutions implemented in a cloud infrastructure. In this paper, we focus on transaction management for such cloud-based graph databases. In particular, we use various graph databases as case studies to survey the different levels of transaction support and concurrency control protocols offered. We also study data distribution issues and replication protocols. Finally, we highlight open issues that need to be addressed in the future.

Keywords: Graph database · Consistency · Cloud computing

1 Introduction

There is an abundance of graph-structured datasets that depict social networks, citation and hyperlink as well as biology, traffic and computer networks. As these datasets grow in volume, efficiently storing and querying them has become an important problem.

There are two main approaches for efficiently managing graph-structured data. The first approach is adopting a data-centric view by using a database management system. In particular, a *graph database* is a database management system that uses operations and queries that expose a graph data model [25]. Thus, graph databases use graph structures with edges, properties and nodes for data storage. They are usually designed to support transactional processing (OnLine Transaction Processing, OLTP) and enforce integrity constraints while providing increased availability as all traditional relational databases. The basic functionality of such graph databases is to support CREATE, READ, UPDATE and DELETE operations on the graph model elements, i.e., nodes, edges and, if available, properties. Some of the most common queries include querying for nodes and edges with particular properties, for the neighbours of a node, for shortest paths in the graph or for particular subgraphs. Popular graph databases

© Springer International Publishing Switzerland 2016
I. Karydis et al. (Eds.): ALGOCLOUD 2015, LNCS 9511, pp. 99–113, 2016.
DOI: 10.1007/978-3-319-29919-8_8

include Neo4j [21], OrientDB [22], Sparksee (formerly known as DEX) [19,29], InfiniteGraph [10], Titan [31] and others.

The second alternative to managing graph structured data is to view graphs as a computational rather than a data model. Systems that adopt this approach are based on parallel processing paradigms, and aim to exploit parallelism to support efficient graph analysis. In contrast to graph databases that focus on OLTP processing, parallel graph processing systems deal with OLAP (OnLine Analytical Processing) analysis, supporting queries that require batch processing, which starts on the whole graph and performs iterative computations. Such algorithms include PageRank and finding cliques or counting triangles in the overall graph. Important representatives include Pregel [18], Giraph [1], GraphLab [17] and PowerGraph [8], Trinity [28], GraphChi [15] and KineoGraph [5].

While parallel graph processing systems consider analysis of mostly static data, a graph database storing graphs corresponding to real networks that evolve over time, needs to provide support for consistent concurrent execution of graph queries and update operations. Thus, transaction support is required in such graph databases. Based on well known graph databases, open source and proprietary, our goal is to study the different approaches graph databases deploy to deal with transactions. We consider both the case of centralized and distributed cloud-based graph databases. In the first case, concurrency control protocols ensure the correct execution of multiple transactions, while in the second case, protocols for replication consistency control and distributed transactions are also needed.

The rest of this paper is structured as follows. In Sect. 2, we briefly present the basic consistency models. Using popular graph databases as case studies, in Sect. 3, we describe data and storage models as well as distribution schemes for the distributed scenario. Section 4 includes the concurrency control protocols deployed and isolation levels achieved by the different databases. Finally, Sect. 5 summarizes our results and highlights open issues and future challenges.

2   Cloud Consistency

Transactions are used in most databases to enable consistent execution of concurrent operations, to provide isolation between users that access the database concurrently and to support recovery from failures while maintaining consistency even in the presence of failures. A *transaction* consists of one or more independent operations (read or write) to a database viewed as one logical unit.

Most databases offer four guarantees namely, *Atomicity, Consistency, Isolation, Durability*, i.e., the well known *ACID* properties.

Atomicity requires that all transactions follow an "all or nothing" rule. If one part of the transaction fails, the entire transaction fails, and the database state is left unchanged.

Consistency ensures that each transactions causes the database to transition from one valid state to another valid state. A valid state is one in which all integrity constraints are satisfied.

Isolation ensures that a transactions does not see the intermediate results of other transactions.

Durability ensures that if a transaction commits, its modifications are not lost.

When the database is distributed, providing high availability is a key requirement. Furthermore, other factors such as the unreliability of the communication network may also influence system performance. As in such distributed settings providing full ACID guarantees induces very large overheads and may be infeasible, other more relaxed consistency models seem to be more appropriate.

Consistency (C) is viewed as equivalent to having a single up-to-date copy of the data, so as that each read operation in the database returns the most recent value of any data item. High-availability (A) means that any request to a non failed node in the system results in a response in a logical time frame. Finally, partition-tolerance (P) ensures that the system remains functional under network partitions. Focusing on these dimensions, Brewer stated the *CAP theorem* [2], that assert that for any distributed system, *"though it's desirable to have consistency, high-availability and partition-tolerance in every system, unfortunately no system can achieve all three at the same time"*.

In particular, it is proved using the asynchronous network model, which assumes a directed graph of processes that use asynchronous (non-blocking) send and synchronous (blocking) receive operations for their communication, that it is impossible to have all three desirable properties, and only two out of the three can be satisfied at the same time [7]. As most distributed systems do not negotiate the requirement of partition-tolerance, the tradeoff that arises is between consistency and availability. However, Brewer returns to the CAP theorem in [3] to emphasize that since network partitions are rather rare, in all other cases a system should offer both perfect availability and consistency.

Based on this tradeoff weaker forms of consistency such as *eventual consistency* [34] have emerged. The eventual consistency model guarantees that if no new updates are made to a data item, eventually all accesses will return the last updated value. Eventual consistency is based on a distributed system paradigm that takes an optimistic approach towards database consistency, while ensuring higher service availability as defined in [23] that defines *Basically Available, Soft state, Eventually consistent* services (BASE).

3   Graph Databases

3.1   Data and Storage Model

Graph databases use graph structures for representing and querying their data, regardless of the underlying storage system. While each database may follow its own specific data model with variations among them, most of them follow an approach similar to the property graph as defined in Neo4j [21,25]. The property graph model consists of nodes, relationships that correspond to the edges between pairs of nodes, and properties contained both in nodes and relationships that are used to describe them. Properties are usually arbitrary key-value pairs.

There are mainly two approaches regarding storage: (i) using a native storage model or (ii) building a graph database on top of another storage model.

Native graph databases are based on a storage model specifically designed for graph structures and optimized to support graph operations. Neo4j [21] uses a native graph storage using separate files to store nodes, relationships and properties. Fixed size records are used both for nodes and relationships, while the latter form a doubly linked list to enable efficient graph traversal. Pointers are also used to link relationships and nodes with lists of their properties. Sparksee [29] takes an object-oriented approach treating nodes and relationships as objects and storing them with corresponding metadata and unique global identifiers.

Graph databases that build upon existing storage models, for instance, serialize a graph structure to store it in a relational database and implement a graph layer supporting the graph operations on top of the relational storage. OrientDB [22] uses a document database and maps the graph model on top of the document model, while InfiniteGraph [10] uses an object-oriented database for storing nodes and relationships as persistent objects. Finally, Titan [31] supports different storage backends: Cassandra [16] and HBase, two key-value storage systems that focus on scalability and distribution and BerkleyDB, a key-value database that offers single-site solutions. The choice of the backend influences the behavior of the graph database with respect to both consistency and scalability issues.

3.2   Distribution Schemes

To improve scalability, load-balance and availability, graph databases are often distributed among multiple sites exploiting some cloud or other similar infrastructure. When distributing a database among sites, one can either *replicate*, *partition* or *shard* the database.

As the three terms are closely related, we first clarify their use in the context of relational databases. Replication refers to creating copies of the entire database or whole tables to more than one site. Data partition encompasses any way of splitting data to different logical parts that are then distributed among multiple sites. In particular, for a relational database, partitioning may refer to either vertical (column-wise) or horizontal (row-wise) partitioning. Finally, sharding refers particularly to horizontal partitioning of a table based on some key value.

For graph databases, we discern between databases that store a single large graph or many smaller graphs. Keeping that in mind, replicating a graph database can either mean replicating the entire database or replicating entire graphs to more than one site. As partition may refer to any type of splitting the data, to avoid confusion we can use it to refer only to distributing different graphs of the same database to different sites. Finally, sharding for graph databases refers to partitioning a single graph among different sites.

When using replication, while availability and load balance is improved, additional consistency issues arise, as the different copies need to be synchronized to provide the most up-to-date information. On the other hand, partitioning and sharding do not involve such consistency concerns, however they may increase the cost of querying and updating the graph as a single query may need information from multiple sites involving increased communication costs.

For sharding, graph traversal becomes expensive as it may require moving from one site to another multiple times. Graph partitioning algorithms that partition (shard) a single graph in different parts that are assigned to different sites aim at reducing the cross site traversals and balancing the load among these sites. Theoretical approaches to construct p-partitions of graphs with the minimum cut, where cut size is defined as the number of edges that connect nodes assigned to different partitions, have been studied in extent (see [6] for a good survey). Recently, new approaches deal with partitioning graph data over the cloud. Most such approaches were developed in the context of graph parallel processing systems to deal with designing an efficient underlying storage model that would be appropriate for the processing model.

Pregel [18] that views a node as the main processing unit, initially used simple hash functions to partition nodes among different sites. As such a simple approach does not consider the structural characteristics of the nodes, improvements have been developed. Mizan [12] not only partitions the nodes of the graph, but also applies corresponding changes and optimizations to the Pregel processing algorithms. An optimizer evaluates the structure of the graph and if the nodes degree distribution follows a power-law, with a small number of nodes having a very large degree while the majority of the nodes have much smaller degrees, then a traditional partitioning algorithm such as METIS [11] is used to derive a partition with a minimum cut for the graph. When the graph does not follow the power law, random partitioning is used instead, as the estimated gains from applying an optimal partition do not justify the cost for its application.

Besides graph properties, the Surfer [4] framework also considers the characteristics of the underlying communication network. The framework recursively partitions with bisection both the data graph, and the network graph which is formed by the sites among which the data will be partitioned. By partitioning both graphs simultaneously, the number of cross-partition edges among data graph partitions is gracefully adapted to the aggregated amount of bandwidth among machine graph partitions. In [33], both graph and network properties are modeled through weights to nodes and edges respectively and a multilevel partitioning algorithm such as [9] is applied outperforming traditional approaches.

For social network graphs, graph properties are exploited along with semantic information in the form of graph communities. In [35], Trinity is used as the graph processing framework and a multi-level label propagation method is deployed. Each node is initially labeled and then labels are updated iteratively as nodes take the label that is prevalent in their neighborhood. The SPAR [24] framework combines partitioning with replication. Partitioning tries to maintain the structure of the communities in the graph, and replication ensures that the neighbors of each node are located at the same partition with the given node. The problem is now formulated as the minimum number of required replicas that satisfies the two requirements. The partitioning algorithm is an incremental greedy local optimization method. In [20], the goal is to reduce the cost involved in SPAR [24] for keeping all the replicas up to date. Not all nodes are replicated and decisions are made locally by monitoring the accesses on each node and using a fairness criterion.

Finally, latest developments include dynamic partitioning and dealing with streaming graph data. One partition scheme may not be appropriate for all graph processing algorithms, since workload balancing in different iterations for each algorithm is possibly different. In [27], dynamic workload balancing is proposed by partitioning the active nodes involved in each graph processing algorithm taking into account the communications required by the steps of the given algorithm. Streaming graphs are considered in [30]. Heuristics are used that try to compute balanced partitions, for instance, by greedily assigning each new node to the partition with the smallest size, or by hashing, or by splitting the input stream in segments and assigning each segment to a different partition.

Despite the recent advances in sharding (partitioning) in the context of graph parallel systems, most graph databases do not inherently support sharding and adopt replication instead. Sharding is delegated as the application's responsibility or dealt with by the backend storage system if such a system exists.

Master-slave replication is the replication scheme most often used. In this scheme, one replica is considered the primary (master), while all other are backups (slaves). The master replica receives all requests and updates are propagated to all slaves that acknowledge the changes. This scheme is used mainly to offer high availability in the presence of failures, as then recovery schemes determine one of the slave replicas as the new master. Multi-master replication allows all replicas to receive and process requests, and updates are propagated to all other replicas. While more efficient, this scheme requires more complex and expensive distributed concurrency control mechanisms.

Sparksee, InfiniteGraph and Neo4j use master-slave replication with synchronous updates and notifications. For instance, Neo4j's distributed solution is called Neo4j HA (High Availability). For a write transaction on a slave, each write is synchronized with the master, while the slave has to be up to date with the master. The transaction first commits on the master and then, if successful, on the slave. Write transactions on the master execute normally as in non-HA mode. If the transaction commits successfully, it is pushed on a number of slaves by using an optimistic protocol. That is, if the push to the slaves fails, the transactions remains successful.

OrientDB, when distributed across a number of clusters, multi master replication is supported within each cluster. All servers are allowed to perform reads and writes on their own replicas, and then notify the rest of the nodes in their cluster, using synchronous or asynchronous notification, increasing consistency in the first case and efficiency in the latter where only eventual consistency is achieved. Synchronization logs are maintained to resolve conflicts and the default policy used gives precedence to older operations. Similarly to the other graph database systems, sharding is not supported by OrientDB.

Titan relies on the backend to handle partition and replication as well. BerkeleyDB is not appropriate for a distributed setting. For the distributed solutions, Cassandra and HBase by default use random partitioning to distribute the nodes among the different servers. However, using an explicit graph partitioning algorithm that tries to minimize the inter-server communication cost is also possible for these two backends.

4   Concurrency Control

With respect to concurrency control, first we discern between graph databases that offer full ACID support and those that on the tradeoff of consistency and availability choose availability and support weaker forms of consistency such as eventual consistency. Non-native graph databases that are build on top of another database or storage system, usually depend on that system and its supported concurrency control. If the backend supports ACID, then the graph database also supports ACID and so on. Databases like Titan, that support different backends may treat transactions and concurrency control differently according to the backend system used each time. Note however that regardless of the approach they follow, most graph databases offer flexibility, allowing them to be configured for weaker consistency when higher availability and efficiency is needed according to the application requirements.

Concurrency controls protocols are divided into *pessimistic* and *optimistic*.

4.1   Pessimistic Concurrency Control

Pessimistic concurrency control blocks any operation of a transaction that may cause a violation of the data integrity, until the danger of the violation no longer exists. The main technique used to block operations is through *locking*. When a transaction needs to access or update a resource, it is first required to acquire a lock on that resource. Only the transaction that possesses the locks on each resource can access or update it. When it completes its operation, the transaction releases the lock so that other transactions can in turn acquire the lock to manipulate the same data. Thus, concurrent access on the same data is controlled and avoided.

Locking protocols may cause deadlocks, if two or more transactions become mutually blocked when waiting for the release of some locks the other ones hold. In this case, complementary techniques for deadlock detection and resolution need to be deployed by the graph database system.

In the context of graph databases, the main difference between locking protocols is *locking granularity*, that is which is the data unit on which locking is acquired. Most graph databases, especially those which deal with workloads consisting of multiple small graphs, require that each transaction that applies any operation on a particular graph should acquire a lock for the entire graph. Thus, the degree of parallelism is reduced since only one transaction per graph is allowed. Graph databases that support more fine granularity in their locking enabling transactions to acquire locks only on particular subsets of nodes and edges support a higher degree of parallelism.

In Neo4j [21], all database operations that access the graph, indexes, or the schema are performed within a transaction. Transactions can be nested as a flat nested transaction, where all nested transactions are added to the scope of a single top level transaction. If a nested transaction aborts or fails for some reason, then the top level transaction is rolled back along with all its other nested transactions. Only write locks are deployed in Neo4j and locking is applied on node and edge level. In particular, when one creates or deletes a node, a write lock is acquired for the particular node. Similarly, to create or delete an edge locks are required for the given edge as well as the nodes that the edge connects. Finally, when adding, deleting or updating properties on nodes or edges, write locks are acquired for the respective nodes or edges. With respect to deadlocks, deadlock detection and manipulation is handled by Neo4j ensuring that when needed the involved transactions will be rolled back and the locks released.

Sparksee [29] supports two types of transactions, read or shared, and write or exclusive, and follows a multiple read/single write model. All transactions unless explicitly declared as write, begin as read transactions and are transformed to write when the first update operation is encountered. For a transaction to be allowed to become a write transaction all other read transactions must finish first. Locking is applied on the graph level, as all manipulations of a graph are encapsulated into sessions that do not allow sharing among multiple threads.

In InfiniteGraph [10], a single graph database is distributed across multiple sites. There is a dedicated lock server process that manages access to the complete database and cooperates with local data server processes that reside in each site hosting a part of the database. Both full and relaxed consistency are supported, where the latter trades off consistency for performance. Locking is performed similarly to Sparksee at a graph level and for full consistency the multiple reads/single write model is deployed. For relaxed consistency, an accelerated form of ingesting nodes is supported, in which the nodes are processed in batches, by non blocking transactions and the database is made eventually consistent. For full consistency, accelerated ingestion may also be supported by secondary processes that handle the coordination of the ingestion.

4.2   Optimistic Concurrency Control

In contrast to pessimistic concurrency control, an optimistic approach operates on the assumption that most likely no conflict will occur during the concurrent execution of the transactions. Thus, it does not block transactions but instead before committing a transaction, it checks whether any integrity constraint has been violated by its execution. In case of a violation, the transaction is rolled back and restarted, otherwise it is committed.

A *multi-version concurrency control (MVCC)* protocol is one of the techniques used for optimistic concurrency control. An MVCC uses timestamps to differentiate between different instances (snapshots) of data items in time. Transactions are also assigned timestamps and determine through their use which version of the database they are reading or writing. The database maintains several versions of each data item which are assigned a write timestamp according to the transaction that has written the item. A transaction can read the most recent version of a data item that precedes the transaction's timestamp. Similarly, each data item is assigned a read transaction corresponding to the latest transaction that has read the item. For a transaction to write an item (a new version), the transaction's timestamp should not precede the read timestamp of the item, and no pending transactions with preceding timestamps should exist. Otherwise, the transaction is aborted and restarted.

OrientDB [22] adopts the MVCC approach for graph databases. Any graph update operation automatically starts a new transaction if no transactions are currently running. Updates made by the transactions are temporary and not visible to other transactions unless the transactions that performed them have committed. Multiple reads and writes are allowed and the versions of the graph elements updated by a transaction are checked upon its commit to determine whether they have been updated by other transactions in which case the first transaction is aborted. When transferring temporary versions to storage, OrientDB additionally acquires exclusive locks to ensure consistency.

4.3   Backend Dependent Concurrency Control

Titan [31], a scalable distributed graph database, supports three different key-value storage backends, Apache Cassandra [16], HBase and BerkeleyDB. Titan relegates transaction management to the backend in use, supporting different consistency guarantees. BerkeleyDB offers limited horizontal scalability and concurrency, on the other hand it is the most appropriate solution for a centralized server scenario. Cassandra and HBase are the best solutions for distributed setups as they provide native support for distributed solutions, but sacrifice consistency or availability.

On BerkeleyDB, transactions are handled by the underlying storage and can be configured to ACID transactions without requiring additional effort from Titan. On eventually consistent storage backends, that is Cassandra and HBase, Titan supports both pessimistic and optimistic concurrency control.

When the pessimistic approach is deployed, Titan must obtain locks in order to ensure consistency because the underlying storage backend does not provide transactional isolation. Titan does not use locking by default, thus, the user has to determine whether locking should be used for each schema element that defines a consistency constraint. The actual lock application mechanism is abstracted such that Titan can use multiple implementations of a locking provider. Currently, it supports two lock implementations: (i) a locking mechanism based on key-consistent reads and writes, which only requires that the backend storage supports key-consistent operations, is supported by both Cassandra and HBase, and implemented based on the use of timestamped lock operations, and (ii) a Cassandra specific locking implementation based on the Astyanax locking recipe, which defines a distributed row lock that performs a sequence of write-read-write operations to effectively lock a row. The recipe also allows for reading the entire row and committing it back as part of the last write. Both locking implementations require that clocks are synchronized across all machines in the cluster.

Optimistic concurrency control is usually preferred in distributed solutions. Since edges are stored as single records in the storage backend, concurrently modifying a single edge leads to conflict. Instead of locking, an edge label can be forked. When modifying an edge whose label is configured to FORK the edge is deleted and the modified edge is added as a new one. Hence, if two concurrent transactions modify the same edge, two modified copies of the edge will exist upon commit which can be resolved during querying traversals if needed.

4.4 Isolation Levels

The highest level of isolation is achieved by *serializability* that ensures that the concurrent execution of multiple transactions is equivalent to some serial execution of them. Serializability incurs very high overheads and considerably limits the degree of concurrent executions. Therefore, most databases relax isolation to lower levels to tradeoff concurrency with efficiency. *Reapatable reads* and *read committed* are usually the isolation levels chosen. With repeatable reads isolation, all reads within the same transaction should read the snapshot established by the first read, while with read committed, all transactions read data that were committed when the query was started.

The supported isolation level in Neo4j is read committed, as a transaction that traverses the graph is not aware of any updates applied by other transactions unless these transactions are committed. This isolation protects transactions from reading dirty data from uncommitted transactions that in the end may be rolled back, but may allow non-repeatable reads as it is not guaranteed that reissuing the same transaction will encounter the same data it has just read. Similarly to Neo4j, OrientDB also supports read committed isolation level.

Compared to Neo4j and OrientDB, Sparksee ensures a higher isolation level, as read and write transactions lock the graph for their duration ensuring serializability. InfiniteGraph also supports a higher isolation level, *snapshot isolation*, which is more relaxed than serializability but still ensures that no uncommitted data is read and no non-repeatable reads allowed.

Finally, Titan when using the BerkeleyDB backend employs as default the repeatable read isolation level. On the other hand, on the backends that do not support transactions inherently, the isolation level is configurable depending on the locking or other concurrency control mechanism deployed.

5 Comparison and Challenges

Tables 1 and 2 summarize the results of our study, focusing on 5 dimensions. Table 1 includes our results regarding the storage model and distribution scheme deployed, and the consistency guarantees offered, while Table 2 presents our results with respect to the concurrency control mechanism used, and isolation level and locking granularity (if applicable) supported.

Regarding the storage model, both native and non native approaches are very popular. There is no clear advantage, and many have argued that even native storage systems are actually based on other storage models such as objects or records, making a comparison between the two options even more difficult.

With respect to the distribution scheme, all databases, besides Titan that relies on Cassandra or HBase and uses their random partitioning scheme, support

Table 1. Database comparison with respect to the storage model, distribution scheme and consistency guarantees.

	Storage	Distribution	Consistency
Neo4j	Native	Master-slave replication	ACID
OrientDB	Document-database	Multi-master replication	ACID
Sparksee	Native	Master-slave replication	ACID
InfiniteGraph	Object-oriented	Synchronous replication	ACID, relaxed
Titan	BerkeleyDB, Cassandra, HBase	Random partitioning	ACID, eventual

Table 2. Database comparison with respect to the concurrency control protocol, isolation level and locking granularity.

	Concurrency control	Isolation level	Granularity
Neo4j	Locking	Read committed	Node/edge
OrientDB	MVCC	Read committed	–
Sparksee	Locking	Serializable	Graph
InfiniteGraph	Locking	Snapshot isolation	Graph
Titan	Locking, optimistic	Repeatable read, configurable	Node/edge

only replication. Thus, graph partitioning is one of the open issues for graph databases as it currently limits their scalability.

With regards to consistency, while most databases claim ACID properties, in reality they all offer configurable consistency, opting for relaxing consistency to improve availability. Titan is the one with the clearer option, enabling the user to select the backend that offers the most for her application, either BerkeleyDB with full ACID or Cassandra and HBase that support eventual consistency but much higher scalability. Neo4j HA is also an option tailored for applications that require higher availability, sacrificing consistency as well.

Locking is the most popular approach with respect to concurrency control. Only OrientDB fully supports optimistic concurrency control with multi-versioning, and Titan when eventual consistency is deployed. As no benchmarks are available, no clear winner can be picked, as depending on the workload, one of the two approaches may be more appropriate.

The isolation level offered by each graph database also varies from read committed to serializable. Serializable is the strictest one offered by Sparksee, but incurring large overheads and decreasing the degree of concurrency. Reapatable reads chosen by Titan in BerkeleyDB offers also high isolation guarantees, while the most popular option is read committed, which is the default option used by most relational databases as well.

Locking granularity significantly influences the concurrency degree of the system. As most systems that deploy locking (except Neo4j and Titan) offer locking on the graph instance level, we consider it a major limitation of current graph databases and another open issue that needs to be addressed.

5.1 Open Issues

We outline next some directions for future research.

Graph Database Partitioning. Though most graph databases offer distributed and cloud-based solutions, they do not support graph partitioning, but instead adopt replication. While replication increases availability and load balance, still it incurs high costs especially for update propagation to maintain consistency. Graph partitioning enables better exploitation of the available resources, and requires less effort for consistency, but with increased communication costs as a query requires data from multiple sites. Though there are many algorithms for finding an optimal graph partitioning scheme that minimizes the nodes cut so as to minimize the number of cross site accesses for a graph traversal, most such algorithms require access to the entire graph in memory and cannot be efficiently applied to the voluminous graphs that are handled by graph databases. Furthermore, as the graphs in graph databases are dynamic, i.e., often updated, a good graph partitioning scheme would need to handle such updates and adapt accordingly. In practice, a combination of replication and partition seems the most promising direction, as queries in graphs often exhibit locality and therefore, nodes in close vicinity should be stored at the same site.

Locking Granularity. When locking is used for concurrency control, most graph databases enable locking at the graph level, i.e., locking an entire graph each time one of its elements is written or read if shared locks are deployed. For databases that consist of workloads of many small graphs this does not seem as a big drawback. However, if the database maintains a few large or even just one large graph, then we can easily discern how limiting this locking mechanism is, restricting any concurrent operations. Neo4j and Titan are the only graph databases that enable locking at a finer granularity, locking only the node(s) and edge elements that are being affected by an update operation. This leads to increased throughput and a greater degree of concurrent transactions. While locking at a finer granularity requires more complex manipulations, it is important for handling large graphs such as social networks and so on, where an update in one node should not forbid any other operation on distant parts of the graph.

Benchmarking. With the wide acceptance of graph databases and the many alternatives which differ from the storage model to the concurrency and replication polices deployed, there is a need for developing a benchmark for comparing the various graph databases with respect to performance. This benchmark, similar to the TPC-C benchmark [32], which simulates a complete computing environment where users issue transactions against a database system, should include scenarios of concurrent transaction executions that combine reads, writes at different levels and measure the transactions per second and other performance metrics, as well as the performance of different database components such as the locking mechanism, deadlock detection and resolution.

Streaming Data. If we consider social networks, and other graph-structured data, an important characteristic is that they are very dynamic in nature, and often show an append-only behavior, where new nodes and edges are added to the graph and existing node and edge properties are updated. This behavior necessitates that graph databases should efficiently handle streaming data. Already, graph databases support node ingestion functionalities [10]. However, more sophisticated operations are needed so that the databases can efficiently handle such streaming data in real time.

Historical Data. As data change, the graph database should maintain older versions of the current graph as valuable information may be mined by observing the evolution of graph data. Thus, graph databases should handle efficient storage of historical graph data to support both historical queries, i.e., queries for older versions of the graph elements that allow the users to query the past, and analysis of historical data to enable users to mine the graph data with applications such as link prediction, graph evolution, and pattern mining for recommendation systems, social networks, but also fraud detection and so on. Little work has been so far focused on dealing with the management of historical data, either proposing storage models outside the context of graph databases [13,14] or dealing with efficient evaluation of specific type of queries [26].

Acknowledgements. Research co-financed by the ESF and Greek national funds through the Operational Program "Education and Lifelong Learning" of NSRF-Research Funding Program: Thales: Cloud9.

References

1. Apache Giraph. http://giraph.apache.org
2. Brewer, E.: Towards robust distributed systems. In: 19th Annual ACM Symposium on Principles of Distributed Computing (Invited Talk), p. 7 (2000)
3. Brewer, E.: CAP twelve years later: how the "rules" have changed. IEEE Comput. **45**(2), 23–29 (2012)
4. Chen, R., Weng, X., He, B., Yang, M., Choi, B., Li, X.: Improving large graph processing on partitioned graphs in the cloud. In: 3rd ACM Symposium on Cloud Computing, Article No. 3 (2012)
5. Cheng, R., Hong, J., Kyrola, A., Miao, Y., Weng, X., Wu, M., Yang, F., Zhou, L., Zhao, F., Chen, E.: Kineograph: taking the pulse of a fast-changing and connected world. In: 7th ACM European Conference on Computer Systems (EuroSys), pp. 85–98 (2012)
6. Fjallstrom, P.O.: Algorithms for graph partitioning: a survey. Linkoping Electron. Art. Comput. Inf. Sci. **3**(10), 1–37 (1998)
7. Gilbert, S., Lynch, N.: Brewer's conjecture and the feasibility of consistent, available. SIGACT News Partition-tolerant Web Serv. **33**(2), 51–59 (2002)
8. Gonzalez, J.E., Low, Y., Gu, H., Bickson, D., Guestrin, C.: Powergraph: distributed graph-parallel computation on natural graphs. In: 10th USENIX Conference on Operating Systems Design and Implementation (OSDI), pp. 17–30 (2012)

9. Hendrickson, B., Leland, P.: A multilevel algorithm for partitioning graphs. In: ACM/IEEE Supercomputing Conference, Article No. 28 (1995)
10. InfiniteGraph. http://www.objectivity.com/infinitegraph
11. Karypis, G., Kumar, V.: Multilevel k-way hypergraph partitioning. In: 36th ACM/IEEE Conference on Design Automation, pp. 343–348 (1999)
12. Khayyat, Z., Awara, K., Alonazi, A., Jamjoom, H., Williams, D., Kalnis, P.: Mizan: a system for dynamic load balancing in large-scale graph processing. In: 8th ACM European Conference on Computer Systems (EuroSys), pp. 169–182 (2013)
13. Khurana, U., Deshpande, A.: Efficient snapshot retrieval over historical graph data. In: 29th IEEE International Conference on Data Engineering (ICDE), pp. 997–1008 (2013)
14. Koloniari, G., Pitoura, E.: Partial view selection for evolving social graphs. In: 1st International Workshop on Graph Data Management Experiences and Systems (GRADES), Article No. 9 (2013)
15. Kyrola, A., Blelloch, G., Guestrin, C.: GraphChi: large-scale graph computation on just a PC. In: 10th USENIX Symposium on Operating Systems Design and Implementation (OSDI), pp. 31–46 (2012)
16. Lakshman, A., Malik, P.: Cassandra - a decentralized structured storage system. ACM SIGOPS Operating Syst. Rev. **44**(2), 35–40 (2010)
17. Low, Y., Bickson, D., Gonzalez, J., Guestrin, C., Kyrola, A., Hellerstein, J.M.: Distributed GraphLab: a framework for machine learning and data mining in the cloud. PVLDB **5**(8), 716–727 (2012)
18. Malewicz, G., Austern, M.H., Bik, A.J., Denhert, J.C., Horn, I., Leiser, N., Czajkowski, G.: Pregel: a system for large-scale graph processing. In: 2010 ACM SIGMOD International Conference on Management of Data, pp. 135–146 (2010)
19. Martinez-Bazan, N., Muntés-Mulero, V., Gómez-Villamor, S., Nin, J., Sánchez-Martinez, M.A., Larriba-Pey, J.L.: DEX: high-performance exploration on large graphs for information retrieval. In: 16th ACM Conference on Information and Knowledge Management (SIGMOD), pp. 573–582 (2007)
20. Mondal, J., Deshpande, A.: Managing large dynamic graphs efficiently. In: 2012 ACM SIGMOD Conference on Information and Knowledge Management, pp. 145–156 (2012)
21. Neo4j. http://neo4j.com/
22. OrientDB. http://orientdb.com/
23. Pritchett, D.: Base: an acid alternative. ACM Queue **6**(3), 48–55 (2008)
24. Pujol, J.M., Erramilli, V., Siganos, G., Yang, X., Laoutaris, N., Chhabra, P., Rodriguez, P.: The little engine(s) that could: scaling online social networks. In: ACM SIGCOMM 2010 Conference, pp. 375–386 (2010)
25. Robinson, I., Webber, J., Eifrem, E.: Graph Databases. O'Reilly, Sebastopol (2013)
26. Semertzidis, K., Pitoura, E., Lillis, K.: TimeReach: historical reachability queries on evolving graphs. In: 18th International Conference on Extending Database Technology (EDBT), pp. 121–132 (2015)
27. Shang, Z., Yu, J.X.: Catch the wind: graph workload balancing on cloud. In: 29th IEEE International Conference on Data Engineering (ICDE), pp. 553–564 (2013)
28. Shao, B., Wang, H., Li, Y.: Trinity: a distributed graph engine on a memory cloud. In: 2013 ACM SIGMOD International Conference on Management of Data, pp. 505–516 (2013)
29. Sparksee. http://www.sparsity-technologies.com/
30. Stanton, I., Kliot, G.: Streaming graph partitioning for large distributed graphs. In: 18th ACM SIGKDD International Conference on Knowledge Discovery and Data Mining, pp. 1222–1230 (2012)

31. Titan. http://thinkaurelius.github.io/titan/
32. TPC Benchmark. http://www.tpc.org/
33. Verbelen, T., Stevens, T., De Turck, F., Dhoedt, B.: Graph partitioning algorithms for optimizing software deployment in mobile cloud computing. J. Future Gener. Comput. Syst. **29**(2), 451–459 (2013)
34. Vogels, W.: Eventually consistent. Commun. ACM **52**(1), 40–44 (2009)
35. Wang, L., Xiao, Y., Shao, B., Wang, H.: How to partition a billion-node graph. In: IEEE 30th International Conference on Data Engineering (ICDE), pp. 568–579 (2014)

Convex Polygon Planar Range Queries on the Cloud: Grid vs. Angle-Based Partitioning

Nikolaos Nodarakis[1]([✉]), Spyros Sioutas[2], Panagiotis Gerolymatos[2], Athanasios Tsakalidis[1], and Giannis Tzimas[3]

[1] Computer Engineering and Informatics Department,
University of Patras, 26500 Patras, Greece
{nodarakis,tsak}@ceid.upatras.gr
[2] Department of Informatics, Ionian University, 49100 Corfu, Greece
{sioutas,c10gero}@ionio.gr
[3] Computer and Informatics Engineering Department,
Technological Educational Institute of Western Greece, 26334 Patras, Greece
tzimas@cti.gr

Abstract. The polygon retrieval problem is, in essence, the problem of preprocessing a set of n 2-dimensional points, so than given a special *ContainedIn* spatial query, the subset of points falling inside the polygon can be reported efficiently. Such queries find great applicability in areas such as computer graphics, spatial databases and GIS applications. However, as the size of spatial data grows rapidly existing centralized solutions fail to retrieve the results in reasonable response time. In this paper, we propose a novel MapReduce algorithm for efficiently processing convex polygon planar range queries in a distributed manner. We apply a grid-based and an angle-based partitioning scheme on the data space and perform a comparative analysis. Through our experimental evaluation we prove that our system is efficient, robust and scalable.

Keywords: Angle · Big data · Convex polygon · MapReduce · Grid · Hadoop · Range queries · Space partitioning

1 Introduction

During recent years, the advances in mobile computing technologies and the massive use of social media has led to an explosive proliferation of available spatial data. Most smartphones carry location positioning equipment (e.g. GPS) and users tend to disseminate their own information to widespread Location Based Services (LBS), such as Facebook, Twitter etc.

A spatial database is a database that manages space information. A spatial query attempts to retrieve information from the database based on certain parameters. An exemplary spatial query is the k-nearest neighbor (kNN) query [4] and its variation aggregated k-nearest neighbor (AkNN) query [19]. However, according to the type of the application and the purpose of use, the analysis may

© Springer International Publishing Switzerland 2016
I. Karydis et al. (Eds.): ALGOCLOUD 2015, LNCS 9511, pp. 114–125, 2016.
DOI: 10.1007/978-3-319-29919-8_9

cover principles like range queries, similarity queries and common path searching, points of interest data mining, etc. In the context of this work, we deal with
range queries over a given geographical two-dimensional region. The problem is
motivated by real life applications in traffic monitoring, cellular communications,
intelligent transportation systems and other domains.

Many centralized algorithms and data structures have been proposed in order
to process such queries efficiently (e.g. R-trees [6] and their variants). But, as
stated and above, the spatial data grow at an exponential rate. The shortcom-
ing of existing centralized solutions is that they cannot be applied any more, as
the resources needed to process such queries exceed the capabilities of a single
server. Consequently, high scalable implementations are required. Cloud computing technologies provide tools and infrastructure to create such solutions and
manage the input data in a distributed way among multiple servers. The most
popular and notably efficient tool is the *MapReduce* [3] programming model,
developed by Google, for processing large-scale data.

In this paper, we propose a novel MapReduce algorithm for efficiently
processing convex polygon planar range queries in a distributed manner in
Hadoop [13,16], the open source MapReduce implementation. Our algorithm
depends on space partitioning on the data space in order to divide equally the
workload among the servers. We examine two schemes, the grid-based and the
angle-based, and compare their performance. To our best knowledge, we are the
first to study the process of convex polygon planar range queries under this
perspective.

The rest of the paper is organised as follows: Sect. 2 presents related work.
In Sect. 3 the preliminaries are presented and the problem definition is stated.
Section 4 pictures the partitioning schemes and describes the MapReduce algorithm. In Sect. 5, we proceed to the experimental evaluation and provide the
outcomes. Finally, Sect. 6 concludes the paper and sketches future research directions.

2   Related Work

The literature of existing solutions in the field of location-dependent queries is
studied extensively by Ilarri et al. [7]. Range queries report the objects within
a specific distance range/region [14,17] and can be static or moving. Canonical polygon range queries have been studied in [12]. The authors present two
approaches that require less space and time compared to the best previous solutions for the general polygon retrieval problem [10]. Moreover, in [11] they propose a method to transform the difficult problem of reporting the trajectory-
lines, which are intersected by a rectangle region, to the simpler one of reporting
points, which lie inside a canonical planar polygon (4-sides) region. To achieve
this, they make use of specific duality transformation methods.

Quite a few techniques and frameworks have been proposed to process spatial
queries on top of MapReduce. The solution presented in [8] process high selec-
tivity queries in HDFS and is based on popular spatial indices such as the R-tree

and its variants. One of the most popular operators in databases is the k nearest neighbor join (kNN join) and has been studied extensively on the MapReduce framework [9,18]. In addition, the frameworks presented in [2,5] support a wide diversity of massive large scale spatial queries on top of MapReduce. Both solutions utilize various indexing techniques to achieve efficient query processing.

In this paper, we propose a novel MapReduce algorithm for efficiently processing convex polygon planar range queries in a distributed manner based on space decomposition techniques. There are no restrictions considering the shape of the polygon, unlike [12].

3   Preliminaries

In this section, at first we define some notation and provide some definitions used throughout this paper and outline the MapReduce model. Then, we circumscribe the halfspace range searching problem and define formally the problem we tackle in the context of this work.

We consider points in a 2-dimensional metric space D defined by a set of 2 dimensions $\{d_1, d_2\}$. Given a dataset P on D with cardinality n, a point $p \in P$ can be represented as $p = \{p_1, p_2\}$ where p_i is the value on dimension d_i and $-\infty \leq p_i \leq \infty$. Moreover, we define $R = \{r_1, r_2, ..., r_k\}$ the k-vertex convex polygon query on D, where r_i is the i-th vertex of R. Given a dataset P and a convex polygon R, a *ContainedIn Query* returns the set of points $CIR_P \subseteq P$ which are contained in R.

Let us further assume that the dataset P is horizontally distributed to N partitions, based on the partitioning technique, such that P_i is the set of points belonging to partition D_i where $P_i \subseteq P$, $\cup_{1 \leq i \leq N} P_i = P$ and $P_i \cap P_j = \varnothing, \forall i \neq j$. Consequently, a point $p \in P$ belongs in CIR_P if and only if there exists a partition $P_i (1 \leq i \leq N)$ with $p \in P_i \subseteq P$ and $p \in CIR_{P_i}$. In other words, we distribute dataset P among the Map tasks (see below) that compute the P_i sets and in the Reduce phase (see below) we compute the local CIR_{P_i} sets, the union of which consists the global CIR_P set. For a complete reference to the symbols used in this paper please see Table 1.

3.1   MapReduce Model

Here, we briefly describe the MapReduce model [3]. The data processing in MapReduce is based on input data partitioning; the partitioned data is executed by a number of tasks executed in many distributed nodes. There exist two

Table 1. Symbols and their meanings

P	Dataset	r_i	i-th vertex of R	h	Halfspace query hyperplane
d	Dataset dimensionality	D_i	i-th data space partition	ϕ_i	i-th angular coordinate
n	Dataset cardinality	P_i	Points of i-th partition	b_i	i-th cartesian coordinate
N	Number of partitions	p_i	i-th coordinate of point p	M_T	Total number of Map tasks
R	Convex polygon query	CIR_{P_i}	Subset of P_i that lay inside R	R_T	Total number of Reduce tasks

major task categories called *Map* and *Reduce* respectively. Given input data, a *Map* function processes the data and outputs key-value pairs. Based on the Shuffle process, key-value pairs are grouped and then each group is sent to the corresponding Reduce task. A user can define his own Map and Reduce functions depending on the purpose of his application. The input and output formats of these functions are simplified as key-value pairs. Using this generic interface, the user can focus on his own problem and does not have to care how the program is executed over the distributed nodes.

3.2 Halfspace Range Searching

The halfspace range searching problem [1] is defined as follows: Preprocess a set S of m points in $\mathbb{R}^d$ into a data structure so that all points satisfying a query constraint $a \cdot x \leq b$ can be reported efficiently. Note that a query corresponds to reporting all points below a query hyperplane h defined by $a \cdot x = b$.

3.3 Problem Statement

In this paper, we decompose the data space into N partitions and assign dataset P to the Map tasks. They compute the local P_i sets and send the output to the Reduce tasks that undertake the process to compute the local CIR_{P_i} sets. Therefore, we focus on the partitioning scheme and the efficient processing of halfspace range searching queries that produce these CIR_{P_i} sets. The formal definition of the problem we tackle in the context of this work follows.

Definition 1. *Problem Definition:* *Given a dataset P in 2-dimensional a data space D, a convex polygon $R = \{r_1, r_2, ..., r_k\}$ and an integer number N, apply a partitioning scheme on D to create N partitions and determine an effective way to process halfspace range searching queries (in each partition) to support efficient convex polygon planar queries in a distributed fashion.*

4 MapReduce Convex Polygon Query Processing

In this section, we introduce the partitioning schemes, describe the map reduce algorithm and proceed in time and space complexity analysis. The angular partitioning scheme follows the principles of the space decomposition technique presented in [15]. Here, we briefly describe the basic idea.

4.1 Hyperspherical Coordinates

We can map the cartesian coordinates of a d-dimensional point x to hyperspherical coordinates, that consist of a radial coordinate v and $d-1$ angular coordinates $\phi_1, \phi_2, ..., \phi_{d-1}$. The transformation is computed using the following equations: $v = \sqrt{x_d^2 + x_{d-1}^2 + ... + x_1^2}$, $\tan \phi_1 = \frac{\sqrt{x_d^2 + x_{d-1}^2 + ... + x_2^2}}{x_1}$, ..., $\tan \phi_{d-2} = \frac{\sqrt{x_d^2 + x_{d-1}^2}}{x_{d-2}}$, $\tan \phi_{d-1} = \frac{x_d}{x_{d-1}}$.

Note that generally $0 \leq \phi_i \leq \pi$ for $i < d-1$, and $0 \leq \phi_{d-1} \leq 2\pi$. However, in our case $d = 2$, $\tan \phi_1 = \frac{x_2}{x_1}$ and $0 \leq \phi_1 \leq 2\pi$.

4.2 Hyperspherical Partitioning

Given the number of partitions N and a d-dimensional space D, the angle-based partitioning scheme assigns to each partition a part D_i of the data space ($1 \leq i \leq N$), where $D_i = [\phi_1^{i-1}, \phi_1^i] \times ... \times [\phi_{d-1}^{i-1}, \phi_{d-1}^i]$, $\phi_j^0 = 0$ and $\phi_j^N = 2\pi$. The symbols ϕ_j^{i-1}, ϕ_j^i declare the boundaries on the angular coordinate ϕ_j for the partition i. Observe that in our case, $D_i = [\phi_1^{i-1}, \phi_1^i]$ since $d = 2$. To derive the boundaries we distinguish two cases according to the distribution of D. In the first case we assume a uniform data distribution. Let V_d be the volume of D, then the volume of each partition should be $\frac{V_d}{N}$. The volume V_d^i of the data space that is projected in the i-th partition is defined as $V_d^i = \int_0^v \int_{\phi_1^{i-1}}^{\phi_1^i} ... \int_{\phi_{d-1}^{i-1}}^{\phi_{d-1}^i} v^{d-1} \sin^{d-2} \phi_1 ... \sin \phi_{d-2} dv d\phi_1 ... d\phi_{d-1}$. Again, since $d = 2$ we have that $V_d^i = \int_0^v \int_{\phi_1^{i-1}}^{\phi_1^i} v dv d\phi_1$.

In the case of non-uniform distributions, we alleviate the problem by choosing a suitable partitioning strategy according to the distribution of the dataset. To derive such an outcome, we define a limit $n_{max} = 2 * n/N$ on the number of points a partition can handle. Initially, only one partition exists and points are assigned to the partition. When the limit is reached, the partition is split in two partitions in a way that both of them contain the same number of points. We continue analogously until all points have been assigned to a partition. For more details please refer to [15].

4.3 Hypercube Partitioning

Respectively, we can define the hypercube partitioning. Given the number of partitions N and a d-dimensional space $D = [-L, L]^d$, the grid-based partitioning scheme assigns to each partition a part D_i of the data space ($1 \leq i \leq N$), where $D_i = [b_1^{i-1}, b_1^i] \times ... \times [b_d^{i-1}, b_d^i]$, $b_j^0 = -L$ and $b_j^N = L$. The symbols b_j^{i-1}, b_j^i declare the boundaries on the cartesian coordinate b_j for the partition i. In this paper we consider 2-dimensional points, so $D_i = [b_1^{i-1}, b_1^i] \times [b_2^{i-1}, b_2^i]$.

In case of uniform distributions the volume of each partition should be $\frac{V_d}{N}$, as before. The volume V_d^i is calculated by the equation $V_d^i = \int_{b_1^{i-1}}^{b_1^i} ... \int_{b_d^{i-1}}^{b_d^i} db_1 ... db_d$. Because $d = 2$, we have that $V_d^i = \int_{b_1^{i-1}}^{b_1^i} \int_{b_2^{i-1}}^{b_2^i} db_1 db_2$. In the case of non-uniform distributions, we apply the same strategy we described in Sect. 4.2.

4.4 Grid vs. Angle-Based Scheme

Consider Fig. 1, where P follows a uniform distribution, $N = 12$ and we apply a convex polygon query R. Moreover, we make the assumption that our dataset is defined in the square $[-L, L]^2$. The sets $P_i (1 \leq i \leq N)$ are produced by the Map tasks. Note that more than one hyperplane constrains may be applied inside a partition (the partitions are defined by the dash lines); for example in partition D_{11} in Fig. 1(a) (D_3 in Fig. 1(b)) the Reduce task outputs the solid black points (local CIR_{P_i}) that are located below h_1 and h_2. The union of CIR_{P_i} sets derive the final CIR_P set. Each hyperspherical coordinate ϕ_i is like ϕ_1 shown in Fig. 1(a). Respectively, each cartesian coordinate b_i is like b_1 and b_2 in Fig. 1(b).

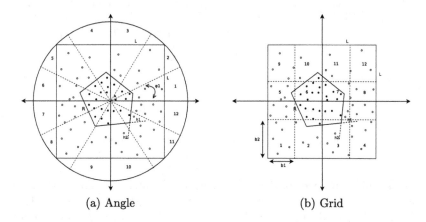

(a) Angle   (b) Grid

Fig. 1. Convex polygon planar range query example

In the case of parallel and distributed processing, where all partitions are examined simultaneously, the performance of grid partitioning may degrade. In this spirit, we propose the angle-based partitioning scheme that can alleviate this problem and diminish the redundant processing. In Fig. 1, the angular partitions are more homogeneous with respect to the query processing and all (only 7 in grid partitioning) have contribution to the global CIR_P set. Therefore, the load balancing of the query process is expected to be evenly distributed among the tasks. However, as seen and by the experiments, in some cases more better load balanced partitions does not mean less processing time.

4.5   Pruning Optimization

In this subsection, we introduce an optimization that greatly enhances the performance of the query processing. As expounded in Sect. 4.4 the Map tasks decide in which partition each point belongs. However, we do not need to check all points of data space whether they are contained in R or not (during Reduce phase). On the contrary, we can prune a significant amount of points thus achieving a huge reduction in the burden of the query process. Consider Fig. 2 in which we display the pruning mechanism in the case of angular space decomposition (the same thing applies for grid space decomposition). If we define the rectangle that encompasses the polygon query R (Fig. 2(b)), we are enabled to prune all points lying outside the rectangle with a simple check. This optimization greatly speeds up the performance of our algorithm as shown by the experimental procedure.

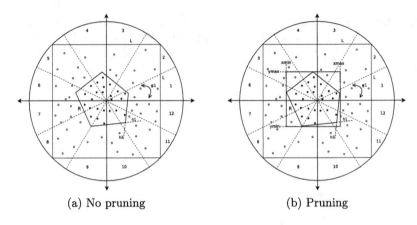

(a) No pruning   (b) Pruning

Fig. 2. Pruning optimization of query processing

Since in this paper we consider 2-dimensional points, we examine the area of the pruned region in the case of $d = 2$ without loss of generality. Given a dataset defined in the hypercube $C = [-L, L]^2$, a convex polygon query R and a rectangle $ER = [x_{min}, x_{max}] \times [y_{min}, y_{max}]$ (Fig. 2(b)) that encloses R we define the pruning optimization PO of ER as $PO(ER) = 100 * \left(1 - \frac{(x_{max} - x_{min}) * (y_{max} - y_{min})}{4L^2}\right)$.

4.6   MapReduce Algorithm

In this subsection, we describe the MapReduce algorithm using pseudo-code and proceed to time and space complexity analysis of each Map and Reduce task. We assume that we have generated the partitions (angle and grid), using the aforementioned equations, during a preprocessing step. At first, we calculate the hyperspherical (cartesian) coordinates $\forall p \in P$ and assign each p to the suitable partition according to the angle-based (grid-based) scheme. Then, we aggregate all points for each set P_i that fall in the same partition D_i and report the local CIR_{P_i}. The Map and Reduce functions are outlined at MapReduce Job 1 pseudo-code that follows.

The *Map* task takes as input the points of P and outputs key-value records, where the key is the partition D_i, in which a point p belongs and the value consists of the point p itself. Then, each *Reduce* task is assigned a partition D_i and for each $p \in P_i$ we apply one or more halfspace range queries. The points that satisfy these queries are added to a list L. The Reduce task emits a key-value pair where the key is the number of partition D_i and the value is list L, i.e. the local CIR_{P_i} set. By combining all CIR_{P_i} sets generated from all Reduce tasks, we yield the overall CIR_P set.

Each Map task runs in $O(n/M_T)$ time, since dataset P is divided horizontally among the total M_T Map tasks. For each Reduce task, assume n_i the number of points that belong to partition D_i and h_i the number of hyperplanes that intersect D_i in the i-th execution of a Reduce function, where $1 \leq i \leq N/R_T$.

MapReduce Job 1

```
 1: function MAP(k1, v1)
 2:     R = {r₁, r₂, ..., rₖ}; , ER = {(x_min, x_max) × (y_min, y_max)};
 3:     coord_vector = computeCoord(v1); Dᵢ = computePart(coord_vector);
 4:     if prune(ER, coord_vector) then \\Prune point if possible
 5:         return;
 6:     end if
 7:     if Dᵢ.intersects(R) then
 8:         output(Dᵢ, v1); \\v1 is a point p = [p₁, p₂]
 9:     end if
10: end function

11: function REDUCE(k2, v2)
12:     L = List{}; hyperplanes = getHPlanes(k2);
13:     for all v ∈ v2 do \\v2 is the set of points in Pᵢ
14:         for all h ∈ hyperplanes do
15:             if !v.satisfies(h) then
16:                 break;
17:             end if
18:         end for
19:         L.add(v);
20:     end for
21:     output(k2, L);
22: end function
```

The time complexity of the Reduce task is clearly affected by the halfspace range query. For each point $p \in P_i$, we can decide if it is located above or below a hyperplane h in constant time $O(1)$. To prove this claim, consider a point $p = \{p_1, p_2\}$ and a hyperplane $h = a \cdot x + b, a < 0, b > 0$. Firstly, we calculate the x_0 and y_0 values where h intersects the x and y axis respectively. Then, we perform the comparisons $p_1 \leq x_0$ and $p_2 \leq y_0$. If they are both true p is located below h, above otherwise. Therefore, each Reduce task needs $O(\sum_i n_i \cdot h_i)$ time to run. The output size is $O(CIR_P)$.

5 Experimental Evaluation

In this section, we conduct a series of experiments to evaluate the performance of our method under many different perspectives. More precisely, we take into consideration the number of partitions N, the size of the dataset and the pruning optimization factor.

Our cluster includes 4 computing nodes (VMs), each one of which has four 2.4 GHz CPU processors, 11.5 GB of memory, 45 GB hard disk and the nodes are connected by 1 GB Ethernet. On each node, we install Ubuntu 14.04 operating system, Java 1.7.0_51 with a 64-bit Server VM, and Hadoop 1.2.1. Moreover, we apply the following changes to the default Hadoop configurations: the replication factor is set to 1; the maximum number of Map and Reduce tasks in each node is set to 3 (consequently we set the number of Reduce tasks to 12), the DFS chunk size is 64 MB and the size of virtual memory for each Map and Reduce task is set to 512 MB.

We assess the approaches CPQMR-Grid and CPQMR-Angle in the experiments, which refer to the aforementioned MapReduce algorithm when applied the grid-based and angle-based partitioning scheme respectively. We evaluate our solution, for both partitioning schemes, using a synthetic dataset of uniformly distributed points. The dataset contains a total of 360,000,000 points and its size is approximately 7.5 GB. We run the experiments against five different polygon queries ($R_1 - R_5$) of increasing size (R_1 is the smallest and R_5 the biggest respectively). Polygons R_1, R_2 and R_4 are of arbitrary shape, polygon R_3 is a triangle and R_5 has a rectangular shape. Moreover, we set L equal to 10,000.

5.1   Effect of Number of Partitions

In this experiment, we test four different configurations for the number of partitions ($N \in \{9, 16, 36, 64\}$) against all polygon queries. The outcome of the experimental process is displayed below in Fig. 3.

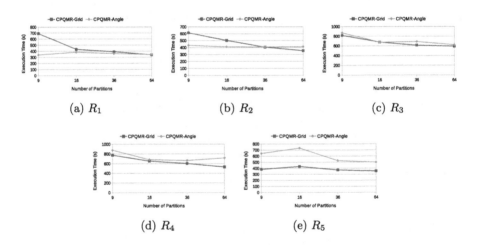

Fig. 3. Different number of partitions

In almost all cases the performance of CPQMR-Grid increases as the number of partitions increments. The same thing does not apply for CPQMR-Angle and this is somewhat expected. Regardless of the number of partitions we decompose the target space, only a fraction of them participates in the query process (i.e. those who intersect R). For $R_1 - R_5$ this fraction is bigger for angle-based partitioning and consider that each partition is assigned in a Reduce task. Moreover, the cluster infrastructure enables us to execute only 12 Reduce tasks simultaneously each time. Thereafter, as the number of partitions grows the performance of CPQMR-Angle either gets worse, either meliorates at a slower pace compared to grid-based scheme. If we could incorporate more nodes to the cluster, CPQMR-Angle would dominate CPQMR-Grid.

Overall, CPQMR-Angle seems to be marginally better for convex polygon queries of arbitrary shape and smaller in size (i.e. R_1 and R_2) as we increase N. If the query area covers a substantial fraction of the data space (R_4), CPQMR-Grid performs better and the curves start to diverge notably for $N > 36$. A very interesting case is the triangular query (R_3) where the two approaches display almost the same behavior and have the worst performance among all polygon query shapes. This happens because the rectangle ER covers a much wider area for R_3 compared to the rest polygon queries. Thus, a lot of redundant points that have no contribution to CIR_P take place in the computation. Finally, we notice that for rectangular queries (R_5) CPQMR-Grid is far better than CPQMR-Angle due to better load balancing between the partitions.

5.2 Effect of Pruning Optimization

In this section, we study the improvement that induces the factor PO to the total query processing for $N = 64$. Due to space limitations we demonstrate the amelioration in performance only for R_1, R_3 and R_5. The results about the enhancement of PO are pictured in Fig. 4. The ratio of improvement in running time fluctuates between 20 % and 50 %. The same thing applies for R_2 and R_4. The above corroborate our claim stated in Sect. 4.5.

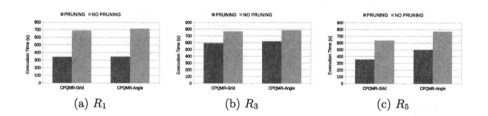

(a) R_1 (b) R_3 (c) R_5

Fig. 4. Pruning optimization efficiency

5.3 Scalability

In the last experiment, we investigate the scalability of the two approaches. We create new chunks smaller in size that are a fraction F of the original dataset, where $F \in \{1/6, 2/6, 3/6, 4/6, 5/6\}$. Moreover, we set the value of N to 9. Figure 5 presents the scalability results yielded for both approaches, which are extremely positive and validate the scalability, robustness and efficiency of our solution. In case of CPQMR-Angle the inference is that the algorithm scales almost linearly as the data size increases. The results for CPQMR-Grid are even better and the behavior of the curve shows a logarithmic tendency for $F > 4/6$.

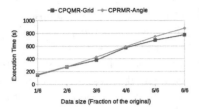

Fig. 5. Scalability

6 Conlusion and Future Steps

In the context of this work we presented a novel MapReduce algorithm for efficiently processing convex polygon planar range queries in a distributed manner. To our best knowledge, we are the first to study convex polygon range queries for large-scale data using an angular partitioning scheme and perform a comparative analysis with the grid partitioning scheme.

In the near future, we plan to perform a more extensive experimental analysis for different data distributions and more dimensions. Moreover, we have in mind to enhance our algorithm using a hybrid grid-angle-based partitioning scheme to achieve even better performance. Finally, we intend to implement our algorithm in other frameworks similar to Hadoop (e.g. Spark) and carry out a comparative analysis between the implementations.

Acknowledgements. This research has been co-financed by the European Union (European Social Fund ESF) and Greek national funds through the Operational Program "Education and Lifelong Learning" of the National Strategic Reference Framework (NSRF) - Research Funding Program: Thales. Investing in knowledge society through the European Social Fund.

References

1. Agarwal, P.K., Arge, L., Erickson, J., Franciosa, P.G., Vitter, J.S.: Efficient searching with linear constraints. In: Proceedings of the 17th ACM SIGACT-SIGMOD-SIGART Symposium on Principles of Database Systems, NY, USA, pp. 169–178. ACM, New York (1998)
2. Aji, A., Wang, F., Vo, H., Lee, R., Liu, Q., Zhang, X., Saltz, J.: Hadoop GIS: a high performance spatial data warehousing system over MapReduce. Proc. VLDB Endow. **6**, 1009–1020 (2013)
3. Dean, J., Ghemawat, S.: MapReduce: simplified data processing on large clusters. In: Proceedings of the 6th Symposium on Operating Systems Design and Implementation, Berkeley, CA, USA, pp. 137–150. USENIX Association (2004)
4. Dunham, M.H.: Data Mining, Introductory and Advanced Topics. Prentice Hall, Upper Saddle River (2002)
5. Eldawy, A.: SpatialHadoop: towards flexible and scalable spatial processing using MapReduce. In: Proceedings of the 2014 SIGMOD Ph.D. Symposium, NY, USA, pp. 46–50. ACM, New York (2014)

6. Guttman, A.: R-trees: a dynamic index structure for spatial searching. In: Proceedings of the 1984 ACM SIGMOD International Conference on Management of Data, NY, USA, pp. 47–57. ACM, New York (2008)
7. Ilarri, S., Mena, E., Illarramendi, A.: Location-dependent query processing: where we are and where we are heading. ACM Comput. Surv. **42**, 12:1–12:73 (2010)
8. Liao, H., Han, J., Fang, J.: Multi-dimensional index on hadoop distributed file system. In: Proceedings of 5th IEEE International Conference on Networking, Architecture, and Storage, pp. 240–249. IEEE Computer Society, Washington, D.C. (2010)
9. Lu, W., Shen, Y., Chen, S., Ooi, B.C.: Efficient processing of k nearest neighbor Joins using MapReduce. Proc. VLDB Endow. **5**, 1016–1027 (2012)
10. Paterson, M.S., Yao, F.F.: Point retrieval for polygons. J. Algorithms **7**, 441–447 (1986)
11. Sioutas, S., Tsakalidis, K., Tsichlas, K., Makris, C., Manolopoulos, Y.: A new approach on indexing mobile objects on the plane. Data Knowl. Eng. **67**, 362–380 (2008)
12. Sioutas, S., Sofotassios, D., Tsichlas, K., Sotiropoulos, D., Vlamos, P.: Canonical polygon queries on the plane: a new approach. J. Comput. **4**, 913–919 (2009)
13. The apache software foundation: Hadoop homepage. http://hadoop.apache.org/
14. Trajcevski, G., Wolfson, O., Hinrichs, K., Chamberlain, S.: Managing uncertainty in moving objects databases. ACM Trans. Database Syst. **29**, 463–507 (2004)
15. Vlachou, A., Doulkeridis, C., Kotidis, Y.: Angle-based space partitioning for efficient parallel skyline computation. In: Proceedings of the 2008 ACM SIGMOD International Conference on Management of Data, NY, USA, pp. 227–238. ACM, New York (2008)
16. White, T.: Hadoop: The Definitive Guide, 3rd edn. O'Reilly Media/Yahoo Press, Sebastopol (2012)
17. Yu, P.S., Chen, S.K., Wu, K.L., Chamberlain, S.: Incremental processing of continual range queries over moving objects. IEEE Trans. Knowl. Data Eng. **18**, 1560–1575 (2006)
18. Zhang, C., Li, F., Jestes, J.: Efficient parallel kNN joins for large data in MapReduce. In: Proceedings of the 15th International Conference on Extending Database Technology, NY, USA, pp. 38–49. ACM, New York (2012)
19. Zhang, J., Mamoulis, N., Papadias, D., Tao, Y.: All-nearest-neighbors queries in spatial databases. In: Proceedings of the 16th International Conference on Scientific and Statistical Database Management, pp. 297–306. IEEE Computer Society, Washington, D.C. (2004)

ART$^+$: A Fault-Tolerant Decentralized Tree Structure with Ultimate Sub-logarithmic Efficiency

Spyros Sioutas[1], Efrosini Sourla[2(✉)], Kostas Tsichlas[3], and Christos Zaroliagis[2,4]

[1] Department of Informatics, Ionian University, 49100 Corfu, Greece
sioutas@ionio.gr
[2] Department of Computer Engineering and Informatics, University of Patras,
26500 Patras, Greece
{sourla,zaro}@ceid.upatras.gr
[3] Department of Informatics, Aristotle University of Thessaloniki,
54124 Thessaloniki, Greece
tsichlas@csd.auth.gr
[4] Computer Technology Institute & Press "Diophantus",
N. Kazantzaki Str., Patras University Campus, 26504 Patras, Greece

Abstract. In this paper, we focus on large-scale, decentralized environments. Our aim is to develop an architecture that can support range queries and scale in terms of number of nodes as well as of data items stored. The solutions proposed in literature are inadequate for our purposes, since their classic logarithmic complexity is too expensive even for single queries. In this work, we contribute the ART$^+$ (Autonomous Range Tree) structure, which outperforms the most popular decentralized structures, since it achieves sub-logarithmic complexity. ART$^+$ achieves an $O(\log_b^2 \log N)$ communication cost for query and update operations, where b is a double-exponentially power of 2 and N is the total number of nodes. Moreover, ART$^+$ is a fully dynamic and fault-tolerant structure, which supports the join/leave node operations in $O(\log \log N)$ expected w.h.p number of hops and performs load-balancing in $O(\log \log N)$ amortized cost. The theoretical performance is verified through experiments.

Keywords: Decentralized systems · Distributed data structures · Load-balancing · Fault tolerance

1 Introduction

Range query processing in decentralized network environments is a notoriously difficult problem to solve efficiently and scalably. It has been studied in the last years extensively, particularly in the realm of P2P, which is increasingly used for content delivery among users. There are many more real-life applications in which the problem also materializes. In cloud infrastructures, a most significant and apparent requirement is the monitoring of thousands of computer nodes,

© Springer International Publishing Switzerland 2016
I. Karydis et al. (Eds.): ALGOCLOUD 2015, LNCS 9511, pp. 126–137, 2016.
DOI: 10.1007/978-3-319-29919-8_10

which often requires support for range queries: consider range queries issued in order to identify under-utilized nodes so as to assign them more tasks, or to identify overloaded nodes so as to avoid bottlenecks in the cloud. For example, we wish to execute range queries such as:

```
SELECT NodeID
FROM CloudNodes
WHERE Low < utilization < High AND os = UNIX
```

Moreover, in cloud infrastructures that support social network services like Facebook, user profiles are stored distributed in several nodes and we wish to retrieve user activity information, executing range queries such as:

```
SELECT COUNT(userID)
FROM CloudNodes
WHERE 3/1/2015 < time < 3/31/2015 AND userID = 123456
      AND NodeID IN Facebook
```

An acceptable solution for processing range queries in such large-scale decentralized environments must scale in terms of the number of nodes as well as in terms of the number of data items stored. The available solutions in literature, are inadequate for our purposes, since for very large volume data (trillions of data items at millions of nodes) the classic logarithmic complexity offered by these solutions, is still too expensive for single queries, not to mention range queries. Further, all available solutions incur large overheads with respect to other critical operations, such as join/leave of nodes, and insertion/deletion of items. Our aim in this work is to provide a solution that is comprehensive and outperforms related work with respect to all major operations, such as lookup, join/leave, insert/delete and load-balancing and to the required routing state that must be maintained in order to support these operations. In particular, we aim at achieving a sub-logarithmic complexity for all the above operations.

In this paper, we contribute the ART$^+$ structure, which outperforms the most popular decentralized structures. The outer level of the proposed structure is an ART[1] structure [11], built by grouping clusters of nodes, whose communication cost of query and update operations is $O(\log_b^2 \log N)$ hops, where the base b is a double-exponentially power of two and N is the total number of nodes. Moreover, ART is a fully dynamic and fault-tolerant structure, which supports the join/leave node operations in $O(\log \log N)$ expected w.h.p number of hops and performs load-balancing in $O(\log \log N)$ amortized cost. Each cluster-node of ART$^+$ is organized as a D^3-Tree[2] [10], which achieves logarithmic bounds for search operations and logarithmic amortized bounds for load-balancing operations. Moreover, D^3-Tree is a highly fault-tolerant structure.

The rest of this paper is organized as follows: Sect. 2 presents related previous work. Section 3 briefly describes the D^3-Tree structure. Our main contribution, the ART$^+$ structure, is described in detail in Sect. 4. Section 5 presents the experimental evaluation. The paper concludes in Sect. 6.

---

[1] Autonomous Range Tree.

[2] Dynamic Deterministic Decentralized Tree.

2 Related Work

Existing structured P2P systems can be classified into two broad categories: Distributed Hash Table (DHT)-based systems and tree-based systems. Examples of the former, which constitute the majority, include Chord, CAN, Pastry, Symphony, Tapestry (see [7] for an overview) and P-Ring [2]. In general, DHT-based systems support exact match queries well and use (successfully) probabilistic methods to distribute the workload among nodes equally. Since hashing destroys the ordering on keys, DHT-based systems typically do not possess the functionality to support straightforwardly range queries, or more complex queries based on data ordering (e.g., nearest-neighbour and string prefix queries). Some efforts towards addressing range queries have been made in [3,8], getting however approximate answers and also making exact searching highly inefficient. The most recent effort towards range queries is the P-Ring [2], a fully distributed structure, that supports both exact match and range queries, achieving $O(\log_d N + k)$ range search performance in average case (N is the number of nodes, d is the $order^3$ of the ring and k is the answer size). It also provides load-balancing, maintaining a load imbalance factor of at most $2 + \epsilon$ in a stable system, for any given constant $\epsilon > 0$, achieving an $O(d \cdot \log_d N)$ performance. P-Ring is considered highly fault-tolerant, using the Chord's Fault Tolerant Algorithms [12].

Tree-based systems are based on hierarchical structures. They support range queries more naturally and efficiently as well as a wider range of operations, since they maintain the ordering of data. However, they lack the simplicity of DHT-based systems, and they do not always guarantee data locality and load balancing in the whole system. Important examples of such systems include Family Trees [7], BATON [5], BATON* [4] and Skip List-based schemes like Skip Graphs (SG), NoN SG, SkipNet (SN), Deterministic SN, Bucket SG, Skip Webs, Rainbow Skip Graphs (RSG) and Strong RSG [7] that use randomized techniques to create and maintain the hierarchical structure. We should emphasize that w.r.t. load-balancing, the solutions provided in the literature are either heuristics, or provide expected bounds under certain assumptions, or amortized bounds but at the expense of increasing the memory size per node. In particular, in BATON [5], the $O(\log N)$ amortized bound of the decentralized overlay (N is the number of nodes in the network) is valid only subject to a probabilistic assumption about the number of nodes taking part in the data migration process, and thus it is in fact an amortized expected bound. Moreover, its successor BATON* [4], exploits the advantages of higher *fanout* (number of children per node), to achieve reduced search cost of $O(\log_m N)$, where m is the *fanout*. However, the higher *fanout* leads to larger update and load-balancing cost of $O(m \cdot \log_m N)$.

Regarding the structures' fault tolerance, BATON [5] maintains vertical and horizontal routing information not only for efficient search, but to offer a large number of alternative paths between two nodes. In its successor BATON* [4], fault tolerance is greatly improved due to higher *fanout*. When $fanout = 2$,

---

[3] Maximum fanout of the hierarchical structure on top of the ring.

approximately 25% of nodes must fail before the structure becomes partitioned, while increasing the fanout up to 10 leads to increasing fault tolerance (60 % of failed nodes partition the structure). However, the cost of load-balancing and updating routing information is greatly increased. A comparison of the afore-mentioned architectures and our proposed structure is given in Table 1.

Table 1. Comparison of P-Ring, BATON*, D^2-Tree, ART and ART$^+$.

Structures	Lookup key	Insert/Delete key (with load-balancing)	Max. size of routing table	Join/Depart node (updating routing tables)
P-Ring	$O(\log_d N)$	$\widetilde{O}(d \cdot \log_d N)$	$O(\log N)$	$\widetilde{O}(d \cdot \log_d N)$
BATON*	$O(\log_m N)$	$\overline{O}(m \cdot \log_m N)$	$O(m \cdot \log_m N)$	$\overline{O}(m \cdot \log_m N)$
D^3-Tree	$O(\log N)$	$\widetilde{O}(\log N)$	$O(\log N)$	$\widetilde{O}(\log N)$
ART	$\widehat{O}(\log_b^2 \log N)$	$\overline{O}(m \cdot \log_m \log N)$	$O(N^{1/4}/\log^c N)$	$\overline{O}(m \cdot \log_m \log N)$
ART$^+$	$\widehat{O}(\log_b^2 \log N)$	$\widetilde{O}(\log \log N)$	$O(N^{1/4}/\log^c N)$	$\widetilde{O}(\log \log N)$

N: number of nodes, d: order of ring, m: fanout, $c > 0$, b: double-exponentially power of 2, $\widehat{O}$: expected bound, $\widetilde{O}$: amortized bound, $\overline{O}$ expected amortized bound.

3   The D^3-Tree Structure

In this section we briefly describe the D^3-Tree structure [10] and present its theoretical background.

The Node Structure: Let N be the number of nodes present in the network and let n denote the size of data ($N \ll n$). The structure consists of two levels. The upper level is a Perfect Binary Tree (PBT) of height $O(\log N)$. The leaves of this tree are *representatives* of the buckets that constitute the lower level of the D^3-Tree. Each bucket is a set of $O(\log N)$ nodes which are structured as a doubly linked list. The structure supports the join/departure operations, while at the same time it tackles failures of nodes whenever these are discovered. Each node v of the D^3-Tree maintains an additional set of links to other nodes apart from the standard links which form the tree:

1. Links to its father and its children.
2. Links to its adjacent nodes based on an in-order traversal of the tree.
3. Links to nodes at the same level as v.
   The links are distributed in exponential steps; the first link points to a node (if there is one) 2^0 positions to the left (right), the second 2^1 positions to the left (right), and the i-th link 2^{i-1} positions to the left (right). These links constitute the *routing table* of v and require $O(\log N)$ space per node.
4. Links to leftmost and rightmost leaf of its subtree. These links accelerate the search process and contribute to the structure's fault tolerance when a considerable number of nodes fail.
5. For leaf nodes only, links to the buckets of the nodes in their routing tables. The first link points to a bucket 2^0 positions left (right), the second 2^1 positions to the left (right) and the i-th link 2^{i-1} positions to the left (right). These links require $O(\log N)$ space per node and keep the structure fault tolerant, since each bucket has multiple links to the main structure.

The next lemma captures some important properties of the routing tables w.r.t. their construction.

Lemma 1. *(i) If a node v contains a link to node u in its routing table, then the parent of v also contains a link to the parent of u, unless u and v have the same father. (ii) If a node v contains a link to node u in its routing table, then the left (right) sibling of v also contains a link to the left (right) sibling of u, unless there are no such nodes. (iii) Every non-leaf node has two adjacent nodes in the in-order traversal, which are leaves.*

Join and Departure of Nodes: The join and departure of nodes may cause the size of the buckets to be uneven, which in the long run renders the structure unbalanced (imagine a bucket holding almost all nodes). To control the size of the buckets a weight-based mechanism is used, which is described in [1], in order to avoid the existence of hotspots.

Redistribution of Nodes: The redistribution guarantees that if there are z nodes in total in the y buckets of the subtree of v, then after the redistribution each bucket maintains either $\lfloor z/y \rfloor$ or $\lfloor z/y \rfloor + 1$ nodes. It also guarantees that each bucket contains $O(\log N)$ nodes, throughout joins or departures of nodes, by employing two operations on the PBT, the *contraction* and the *extension* (Fig. 1). When a redistribution takes place at the root of the PBT, the structure also checks whether any of these two operations can be applied to the PBT. The extension operation adds one more level of nodes at the PBT from existing nodes in the buckets, thus increasing its height by one. The contraction operation removes one level of nodes from the PBT and puts them into the buckets, thus decreasing its height by one. These two operations involve a reconstruction of the structure which rarely happens.

The Index Structure: The range of all values stored in the overlay is partitioned into sub-ranges, each one of which is assigned to a node of the overlay. An internal node v with range $[x_v, x'_v]$ may have a left child u and a right child w with ranges $[x_u, x'_u]$ and $[x_w, x'_w]$ respectively such that $x_u < x'_u < x_v < x'_v < x_w < x'_w$. Ranges are dynamic in the sense that they depend on the values maintained by the node.

Search and Range Queries: The search for an element a in a D^3-Tree of N nodes may be initiated from any node v at level l and is carried out in $O(\log N)$ steps. A range query $[a, b]$ reports all elements x such that $x \in [a, b]$. A range query $[a, b]$ initiated at node v, invokes a search operation for element a. Node u that contains a returns to v all elements in this range and then the range query is forwarded to the right adjacent node (in-order traversal) and continues until an element larger than b is reached for the first time.

Updates and Load-Balancing: Assume that an update operation (insertion/ deletion) is initiated at node v involving element a. By invoking a search operation, node u with range containing element a is located and the update operation is

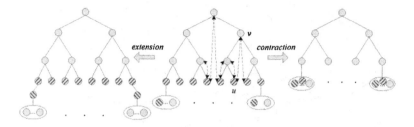

Fig. 1. The initial D^3-Tree structure (middle) and the operations of extension (left) and contraction (right).

performed on u. Afterwards, the weight-based mechanism [1] is applied if necessary, to redistribute elements among nodes.

Fault Tolerance: Searches and updates in the D^3-Tree do not tend to favour any node, and in particular nodes near the root. However, a single node can be easily disconnected from the overlay, when all nodes with which it is connected fail. This means that 4 failures (two adjacent nodes and two children) are enough to disconnect the root. The most easily disconnected nodes are those which are near the root, since their routing tables are small in size.

  If a node v discovers that node u is unreachable, then it contacts a sibling of u through the routing tables of the siblings of v. This sibling of u is able to reconstruct all links of node u and a node departure for u is initiated, which resolves this failure.

Performance: A D^3-Tree structure with N nodes and n data elements residing on them achieves: (i) $O(\log N)$ space per node; (ii) deterministic $O(\log N)$ search cost; (iii) deterministic amortized $O(\log N)$ update cost both for element updates and for node joins and departures; (iv) deterministic amortized $O(\log N)$ bound for load-balancing. The D^3-Tree supports ordered data queries optimally, and tolerates node failures. For evaluation purposes, a simulator was built, with a user friendly interface and a graphical representation of the structure, which is publicly available[4].

4   The ART$^+$ Structure

In this work, we contribute the ART$^+$ structure, which outperforms the most popular decentralized structures of literature. ART$^+$ is similar to its predecessor, ART [11] regarding the structure's outer level. Their difference, which introduces performance enhancements, lies in the fact that each cluster-node of ART$^+$ is structured as a D^3-Tree [10].

Building the ART$^+$ Structure: The backbone structure of ART$^+$ is similar to LRT[5], in which some interventions have been made to improve its performance and increase the robustness of the whole system. ART$^+$ is built by grouping

---

[4] https://github.com/sourlaef/d3-tree-sim.
[5] LRT: Level Range Tree.

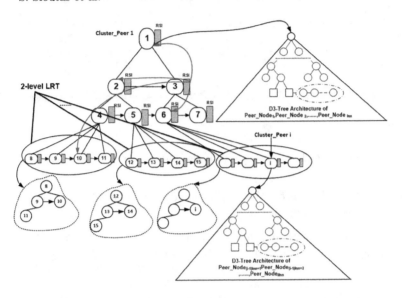

Fig. 2. The ART$^+$ structure for b = 2.

cluster-nodes having the same ancestor and organizing them in a tree structure recursively. A cluster-node is defined as a bucket of ordered nodes. The innermost level of nesting (recursion) will be characterized by having a tree in which no more than b cluster-nodes share the same direct ancestor, where b is a double-exponentially power of two (e.g. 2, 4, 16,...). Thus, multiple independent trees are imposed on the collection of cluster-nodes. The height of ART$^+$ is $O(\log \log_b N)$ in the worst case. The ART$^+$ structure remains unchanged w.h.p. Figure 2 illustrates a simple example, where $b = 2$.

The degree of the cluster-nodes at level $i > 0$ is $d(i) = t(i)$, where $t(i)$ indicates the number of cluster-nodes at level i. It holds that $d(0) = b$ and $t(0) = 1$. At initialization step, the 1st node, the $(\ln n + 1) - th$ node, the $(2 \cdot \ln n + 1) - th$ node and so on are chosen as bucket representatives, according to the balls in bins combinatorial game presented in [6]. Let n be w-bit keys, N be the total number of nodes and N' be the total number of cluster-nodes. Each node with label i (where $1 \leq i \leq N$) of a random cluster, stores ordered keys that belong in the range $[(i - 1) \ln n, i \ln n - 1]$, where $N = n/\ln n$. Each cluster-node with label i' (where $1 \leq i' \leq N'$) stores ordered nodes with sorted keys belonging in the range $[(i' - 1) \ln^2 n, i' \ln^2 n - 1]$, where $N' = n/\ln^2 n$ or $N' = N/\ln n$ is the number of cluster-nodes.

ART$^+$ stores cluster-nodes only, each of which is structured as an independent decentralized architecture, which changes dynamically after node join/leave and element insert/delete operations inside it. In contrast to its predecessor, ART, whose inner level was structured as a BATON*, each cluster-node of ART$^+$ is structured as a D^3-Tree. Each cluster-node is equipped with a routing table named Random Spine Index (RSI), which stores pointers to cluster-nodes belonging to a random spine of the tree (instead of the LSI[6] of LRT which stores pointers to the nodes of the left-most spine, decreasing this way the robustness of the structure). Moreover, instead of using fat CI[7] tables, which store pointers to the collections of nodes presented at the same level, the appropriate collection of cluster-nodes is accessed by using a 2-level LRT structure. In ART⁺, the overlay of cluster-nodes remains unaffected in the expected case w.h.p. when nodes join or leave the network.

Load Balancing: The operation of join/leave of nodes inside a cluster-node is modelled as the combinatorial game of balls in bins presented in [6]. In this way, for an $\mu(\cdot)$ random sequence of join/leave node operations, the load of each cluster node never exceeds $\Theta(\log N)$ size and never becomes zero in expected w.h.p. case. In skew sequences, though, the load of each cluster-node may become $\Theta(N)$ in worst case. The load-balancing mechanism for a D^3-tree structure, as described previously, has an amortized cost of $O(\log K)$, where K is the total number of nodes in the D^3-tree. Thus, in an ART⁺ structure, the cost of load-balancing is $O(\log \log N)$ amortized.

Routing Overhead: We overcome the problem of fat CI tables with routing overhead of $O(\sqrt{N})$ in worst case, using a 2-level LRT structure. The 2-level LRT is an LRT structure over $\log^{2c} Z$ buckets each of which organizes $Z/\log^{2c} Z$ collections in a LRT manner, where Z is the number of collections at current level and c is a big positive constant. As a consequence, the routing information overhead becomes $O(N^{1/4}/\log^c N)$ in the worst case.

Lookup Algorithms: Since the structure's maximum number of nesting levels is $O(\log_b \log N)$ and at each nesting level i we have to apply the standard LRT structure in $N^{1/2^i}$ collections, the whole searching process requires $O(\log_b^2 \log N)$ hops. Then, we have to locate the target node by searching the respective decentralized structure. Through the polylogarithmic load of each cluster node, the total query complexity $O(\log_b^2 \log N)$ follows. Exploiting now the order of keys on each node, range queries require $O(\log_b^2 \log N + |A|)$ hops, where $|A|$ is the answer size.

Join/Leave Operations: A node u can make a join/leave request to a node v, which is located at cluster node W. Since the size of W is bounded by a *polylogN* size in expected w.h.p. case, the node join/leave can be carried out in $O(\log \log N)$ hops. The outer structure of ART⁺ remains unchanged w.h.p. as mentioned before, but each D^3-tree structure changes dynamically after node join/leave operations. According to D^3-Tree performance evaluation, the node join/leave can be carried out in $O(\log \log N)$ hops.

Node Failures and Network Restructuring: In the ART⁺ structure, similarly to ART, the overlay of cluster-nodes remains unchanged in the expected case w.h.p., so in each cluster-node the algorithms for node failure and network restructuring

---

[6] LSI: Left Spine Index.
[7] CI: Collection Index.

are according to the decentralized architecture used. D^3-Tree is a highly fault-tolerant structure, since it supports procedures for node withdrawal and handles massive node failures efficiently.

5 Performance Evaluation

In this section we evaluate the performance of ART$^+$ structure and compare it to the previous structure, ART. Each cluster node of the ART and ART$^+$ is a BATON* and D^3-Tree structure respectively. BATON* was implemented and evaluated in [4], while ART was evaluated in [11], using the Distributed Java D-P2P-Sim simulator presented in [9]. The source code of the whole evaluation process, which showcases the improved performance, scalability, and robustness of ART over BATON* is publicly available[8]. For the performance evaluation of ART$^+$, we used the D^3-Tree simulator.

To evaluate the performance of ART and ART$^+$ for the lookup and load-balancing operations, we ran experiments with different number of total nodes N from 50,000 to 500,000. As proved in [11], each cluster node stores no more than $0.75 \log^2 N$ nodes in smooth distributions (normal, beta, uniform) and no more than $2.5 \log^2 N$ nodes in non-smooth distributions (powlow, zipfian, weibull). Moreover, we inserted elements equal to the network size multiplied by 2000, which are numbers from the universe $[1...1,000,000,000]$. We used the number of passing messages to measure the performance.

Note here that, as proved in [11], ART outperforms BATON* in lookup operations, except for the case where $b = 2$. Moreover, ART achieves better load-balancing compared to BATON*, since the cluster-node overlay remains unaffected w.h.p. through joins/departures of nodes and the load-balancing performance is restricted inside a cluster-node. Consequently, in this work, ART$^+$ is compared directly to ART.

Cost of Lookup Operations. To measure the network performance for the lookup operations (single and range queries), we conducted experiments for different values of b, 2, 4 and 16, in which for each N, we executed 1,000 single queries and 1,000 range queries. The search cost is depicted in Fig. 3. Both normal (beta, uniform) and worst cases (powlow, zipfian, weibull) are depicted in the same graph. Experiments confirm that the query performance of ART and ART$^+$ is $O(\log_b^2 \log N)$ and the slight performance divergences are due to the fact that BATON*, as the inner structure of ART's cluster-node, performs better that D^3-Tree in search operations.

In case of massive failures, the search algorithm has to find alternative paths to overcome the unreachable nodes. Thus, an increase in node failures results in an increase in search costs. To evaluate the system in case of massive failures, we initialized the system with 10,000 nodes and let them randomly fail without recovering. At each step, we check if the network is partitioned or not. Since the backbone of ART and ART$^+$ remains unaffected w.h.p., the search cost

---

[8] http://code.google.com/p/d-p2p-sim/.

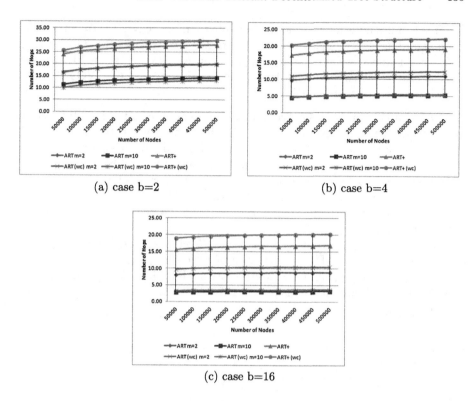

(a) case b=2

(b) case b=4

(c) case b=16

Fig. 3. Cost of lookup operations

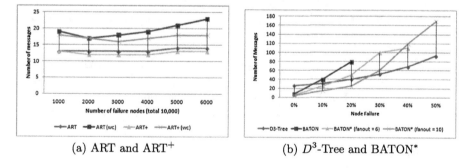

(a) ART and ART$^+$

(b) D^3-Tree and BATON*

Fig. 4. Lookup operations with node failures

is restricted inside a cluster-node (BATON* or D^3-Tree respectively), meaning that b parameter does not affect the overall expected cost. Figure 4a illustrates the effect of massive failures.

We observe that both structures are fault tolerant since the failure percentage has to reach the threshold of 60 % to partition them. Moreover, even in the worst case scenario, the ART$^+$ maintains lower search cost compared to ART, since

D^3-Tree handles node failures more effectively than BATON*. To strengthen our claim regarding the enhanced performance of D^3-Tree towards BATON* in case of massive failures, we present their performances, as depicted in Fig. 4b. We observe that D^3-Tree maintains low search cost, compared to BATON*, even for failure percentage of $\geq 30\%$.

Cost of Load-Balancing Operations. To evaluate the cost of load-balancing, we tested the network with a variety of distributions. For a network of N total nodes, we performed $2N$ node updates. Both ART and ART$^+$ remain unaffected w.h.p., when nodes join or leave the network, thus the load-balancing performance is restricted inside a cluster-node (BATON* or D^3-Tree respectively), meaning that b parameter does not affect the overall expected cost. The load-balancing cost is depicted in Fig. 5a. Both normal and worst cases are depicted in the same graph.

Experiments confirm that ART$^+$ has an $O(\log \log N)$ load-balancing performance, instead of the ART performance of $O(m \cdot \log_m \log N)$. Thus, even in the worst case scenario, the ART$^+$ outperforms ART, since D^3-Tree has a more efficient load-balancing mechanism than BATON* (Fig. 5b).

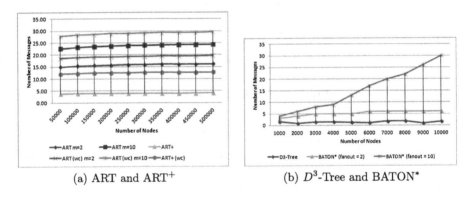

(a) ART and ART$^+$ (b) D^3-Tree and BATON*

Fig. 5. Cost of load-balancing operation

6   Conclusions

In this paper, we presented a new efficient decentralized infrastructure for range query processing with probabilistic guarantees, the ART$^+$ structure. We presented in brief the theoretical algorithmic analysis, which showed that the communication cost of query operations, element update and node join/leave operations scale sub-logarithmically expected w.h.p. Moreover, the cost for the load-balancing operation is sub-logarithmic amortized. Experimental comparison to its predecessor, the ART structure, showed slightly less efficiency towards lookup operations (single and range queries), but improved performance for the load-balancing operation and the lookup operations in case of node failures. Moreover, experiments confirm that ART$^+$ is highly fault-tolerant in case of massive failures. Note that, so far, ART outperforms the state-of-the-art decentralized structures.

Acknowledgments. This research has been co-financed by the European Union (European Social Fund - ESF) and Greek national funds through the Operational Program "Education and Lifelong Learning" of the National Strategic Reference Framework (NSRF) – Research Funding Programs Thales & Heracletus II, Investing in knowledge society through the European Social Fund.

References

1. Brodal, G., Sioutas, S., Tsichlas, K., Zaroliagis, C.: D^2-tree: a new overlay with deterministic bounds. Algorithmica, pp. 1–22, April 2014
2. Crainiceanu, A., Linga, P., Machanavajjhala, A., Gehrke, J., Shanmugasundaram, J.: Load balancing and range queries in p2p systems using p-ring. ACM Trans. Internet Technol. **10**(4), 16: 1–16: 30 (2011)
3. Gupta, A., Agrawal, D., Abbadi, A.E.: Approximate range selection queries in peer-to-peer systems. In: Proceedings of the 1st Biennial Conference on Innovative Data Systems Research (CIDR 2003) (2003)
4. Jagadish, H.V., Ooi, B.C., Tan, K., Vu, Q.H., Zhang, R.: Speeding up search in p2p networks with a multi-way tree structure. In: Proceedings of ACM International Conference on Management of Data (SIGMOD 2006), Chicago, Illinois, USA, pp. 1–12 (2006)
5. Jagadish, H.V., Ooi, B.C., Vu, Q.H.: Baton: a balanced tree structure for peer-to-peer networks. In: Proceedings of the 31st Conference on Very Large Databases (VLDB 2005), Trondheim, Norway, pp. 661–672 (2005)
6. Kaporis, A.C., Makris, C., Sioutas, S., Tsakalidis, A., Tsichlas, K., Zaroliagis, C.D.: Improved bounds for finger search on a RAM. In: Di Battista, G., Zwick, U. (eds.) ESA 2003. LNCS, vol. 2832, pp. 325–336. Springer, Heidelberg (2003)
7. Ozsu, M.T., Valduriez, P.: Principles of Distributed Database Systems. Springer, New York (2011)
8. Sahin, O., Gupta, A., Agrawal, D., Abbadi, A.E.: A peer-to-peer framework for caching range queries. In: Proceedings of the 20th Conference on Data Engineering (ICDE 2004), pp. 165–176. IEEE, March 2004
9. Sioutas, S., Papaloukopoulos, G., Sakkopoulos, E., Tsichlas, K., Manolopoulos, Y.: A novel distributed p2p simulator architecture: D-p2p-sim. In: ACM CIKM, pp. 2069–2070 (2009)
10. Sioutas, S., Sourla, E., Tsichlas, K., Zaroliagis, C.: D^3-Tree: a dynamic deterministic decentralized structure. Algorithms - ESA 2015. LNCS, vol. 9294, pp. 989–1000. Springer, Heidelberg (2015)
11. Sioutas, S., Triantafillou, P., Papaloukopoulos, G., Sakkopoulos, E., Tsichlas, K.: Art: sub-logarithmic decentralized range query processing with probabilistic guarantees. J. Distrib. Parallel Databases (DAPD) **31**(1), 71–109 (2012)
12. Stoica, I., Morris, R., Karger, D., Kaashoek, M.F., Balakrishnan, H.: Chord: a scalable peer-to-peer lookup service for internet applications. SIGCOMM Comput. Commun. Rev. **31**(4), 149–160 (2001)

Comparison of Database and Workload Types Performance in Cloud Environments

George Seriatos, George Kousiouris[✉], Andreas Menychtas,
Dimosthenis Kyriazis, and Theodora Varvarigou

National Technical University of Athens, 9 Heroon Politechniou Street,
15779 Zografou, Greece
{g.seriatos,gkousiou,ameny,dimos,dora}@mail.ntua.gr

Abstract. The rapid growth of unstructured data over the last few years, has led to the emergence of new database management systems. Traditional relational databases, despite their wide adoption and plethora of features, begin to show weaknesses when having to deal with very large amounts of data. Numerous types of databases have emerged in the Cloud domain, in order to exploit the elasticity of Cloud environments, while relaxing the typical ACID considerations and investigating trade-offs of the CAP theorem. The aim of this paper is to investigate how such offerings (MongoDB, Cassandra and HBase namely), based on these tradeoffs, behave when deployed in virtual environments (of the BONFIRE facility) and how they are measured against widely used benchmarks such as YCSB. The results may be helpful for potential adopters to choose from these offerings, based on their individual needs for specific workloads or query structures.

Keywords: Cloud computing · NoSQL · Performance · YCSB · HBase · MongoDB · Cassandra · Benchmarks

1 Introduction

In recent years, the generation and storage of enormous data sizes has led to the creation and investigation of numerous, tailored per case data management systems, that go beyond the typical SQL databases applications. It is estimated that the total data size of the digital universe is equivalent to 2.84 ZB (billion terabytes) with predictions raising this number to 40 ZB for 2020 [1]. Facebook alone gathers 300 PB of data, from which it processes at least 1 PB per month [2]. The Large Hadron Colider in CERN gathers approximately 15 PB per year for processing [3]. The term Big Data is used for one of the most emerging technologies in order to describe the concentration, storage and analysis of especially large data volumes for the extraction of conclusions, correlations and trends. Areas that are affected by this analysis include meteorology, genomics, market analysis among others.

Data management solutions can considerably benefit from their deployment and instantiation in cloud computing environments [16], however their performance may often differ, significantly in many cases depending on the configuration and offering,

© Springer International Publishing Switzerland 2016
I. Karydis et al. (Eds.): ALGOCLOUD 2015, LNCS 9511, pp. 138–150, 2016.
DOI: 10.1007/978-3-319-29919-8_11

which affects the operational aspects of each solution so as to effectively exploit the cloud computing innovations [17]. However, the operation in an abstracted and distributed environment introduces also significant challenges e.g. the CAP theorem [18] which states that you can obtain at most two out of three properties: Consistency, Availability and tolerance to network Partitions. In large scale data management approaches where data are replicated and distributed, consistency is compromised in order provide high availability, thus relaxing the ACID guarantees (Atomicity, Consistency, Isolation, Durability) of the system for database transactions [19].

Traditional relational databases portray a number of significant weaknesses in the analysis of these data, that may origin either from their sheer volume or from the fact that in many cases they follow unstructured data formats that are difficult to be translated into rigid database schemas. While the SQL solutions are oriented towards aspects such as consistency and concurrent transactions, they fall short in use cases where large partitioning or availability needs are necessary. This gap has started to be covered in recent years through the development of NoSQL systems that tend to abandon some of the typical characteristics of relational databases (such as ACID characteristics) in order to ensure the ability for parallel and distributed storage and processing without structural constraints (e.g. table structures).

Development of NoSQL systems has been extremely rapid in recent years. Currently there are 150 available systems [4]. The separation in categories is performed mainly through their capabilities or through the way they store data. An indicative categorization of NoSQL systems appears in [5]. The main purpose of this paper is to investigate a number of such solutions (namely MongoDB, HBase and Cassandra) in a variety of usage scenarios, based on different types of workloads, and extract a number of measurements and conclusions with relation to each system's ability to handle the respective traffic. To this end, the YCSB benchmark client [6] is used in order to launch queries against deployed system instances in the Bonfire experimental Cloud platform [7]. The paper is structured as follows. In Sect. 2 related work in the respective field of Cloud and DB benchmarking is presented, while in Sect. 3 the key characteristics of the selected databases are presented, along with information on the automation and setup of the measurement process. Section 4 presents the performed experiments and measurements results while Sect. 5 concludes the paper.

2 Related Work

There are several approaches that analyze the performance aspects of the database management solutions in cloud. In RDBMS, the TPC benchmarks were the most prominent benchmarking approaches for measuring the performance characteristics [10]. [11] conducted benchmarks on several existing cloudbased management systems: (a) data read and write benchmark with seven tasks to evaluate the read and write performance in different situations, and (b) structured query benchmark focusing on basic operations in the structured query language such as key words matching, range query and aggregation.

NoSQL cloud management systems can be categorized as key-value stores, document stores, column stores and graph stores [12]. Their characteristics include a variety of different data models (fully structured semi structured and unstructured), different querying and most importantly different scaling methodologies to support data partitioning, replication, consistency and concurrent access. These characteristics, the execution environment configuration, have immense effect on the performance of the data management operations, positive or negative. Brian F. Cooper et al. propose YCSB framework [6] that facilitates performance comparisons of modern cloud data serving systems and define a core set of benchmarks and report results for four widely used systems: Cassandra, HBase, Yahoo!'s PNUTS and MySQL. Their analysis examines the aspects of (a) read performance versus write performance, (b) latency versus durability, (c) synchronous versus asynchronous replication and (d) data partitioning, in operation of read, update, scan and insert.

Authors in [13] argue that standardized performance benchmarking is required so as to evaluate the eventual consistency in distributed key-value storage systems and propose a methodology that extends the popular YCSB benchmark to measure the staleness of data returned by reads using the concept of Δ-atomicity [14]. [15] presents an evaluation of the performance for database management, SQL and NoSQL, in the domain of IoT and particularly for sensor data. Besides the difference in performance between SQL and NoSQL solutions, their analysis results show a considerable impact on the performance when the databases are deployed in virtualized cloud environments. In most cases the impact is negative however, only a specific deployment has been tested and no other cloud offerings and/or configurations are examined.

3   DB Features and Measurement Automation

3.1   DB Features

With regard to the selected databases, the goal was to differentiate between features and strategies of available systems. Thus in terms of architecture, HBase follows a more centralized master slave approach, while Cassandra a more peer to peer one. HBase is written in Java and is tightly integrated with the underlying file system (HDFS) and with the MapReduce Apache Hadoop framework and can offer consistency guarantees. It follows a key value approach and stores data in a column oriented format on disk. It also offers atomicity of operations on a row level. Cassandra is also written in Java and the main difference is that it does not portray a single point of failure. In order to enroll a node in the system, one only needs to start the basic daemon process and insert information on one existing system node. In practice a number of nodes are determined as seeds and they are the ones that undertake the role of enrolling a new node in the ring. Cassandra's main benefit is the lack of one master node, which improves system resiliency and availability, however it comes with a price on the consistency levels achieved. It follows a column oriented data organization. One extra feature with relation to HBase is the ability to define composite column families that can serve as an extra layer of organization. Thus it may depict concentrated data from multiple columns. Cassandra also offers atomicity on a row level, however contrary to HBase it can not offer atomicity in cases of updating more than one rows in a single transaction. Different consistency levels are offered, that are coupled with the used replication factor and they regulate the need to have consistent replicas across the system.

MongoDB on the other hand is a document oriented DB, meaning that it stores the data in fields and can be directly queried based on their contents. Data are stored in the form of BSON (Binary JSON). It may also support secondary indexes for faster search. The replica sets in Mongo are defined as primary or secondary. Update of the secondary instances is performed synchronously, affecting latency, by adjusting the "write concern" option or asynchronously, in order not to bottleneck the system. However the latter has an effect on consistency for read operations from them. Based on the consistency option selection this read operation may be limited only to the primary copy. A major difference of this DB is the fact that it mainly uses memory-mapped files in order to enhance performance.

3.2   Cloud Facility Setup

The experiments were performed in the BonFIRE Cloud computing testbed [7]. The purpose of this facility is to provide an experimental facility for Cloud Computing research, across various locations and heterogeneous resources. Management of the available resources is performed through OCCI [8], in order to offer a homogeneous interface with heterogeneous infrastructures (OpenNebula, Virtual Wall, Cells). For the purposes of the experiment, the resources in Table 1 were used. Node 1 was selected with increased capabilities in order to serve as the Master node in HBase and HDFS and it was enriched with more functionalities in the other two systems as well. OS in all cases was Debian Squeeze v6 (kernel:2.6.32-5-amd64).

HBase was the system that presented the most challenges in terms of setup, since in many cases the errors occurred were not adequately described. Special care was given to the networking setup (especially to restrict the use of IPv6), and to remove the loopback address since it caused connection errors to the other nodes. It was also the system that appeared to be more affected by the RAM shortage in the available nodes.

Cassandra was the system that was easier to manage and configure, through the configuration of seed node details. In the case of Mongo, a series of manual steps were necessary. In order to ensure a replication factor of 3, 3 mongod processes need to start, that are configured with relation to which one is the primary, in order to kick off the different shards.

The distribution of the replicas followed the logic of a Cassandra ring. Ext4 file system was used for the main data since it is considered more efficient. One characteristic of Mongo is that due to the preallocation techniques used, it requires significantly higher disk space to start, in comparison to the actual data stored. Given that the amount of disk space was limited, a number of options needed to be utilized during the configuration that limit the initial size of the DB ("–smallfiles" and "–oplogSize 128").

Table 1. BonFIRE resources used

	Node 1	Nodes 2–6	Node 7
CPU	4 Physical Cores (AMD Opteron 6176)	2 Physical Cores (2:AMD Opteron 6176, 3–6: Intel Xeon E5620)	4 Physical Cores (AMD Opteron 6176)
RAM	10 GB	4 GB	1 GB
Disk	10 GB ext3: OS + Software	10 GB ext3: OS + Software	10 GB ext3: OS + Software
	10 GB ext4: Data	10 GB ext4: Data	
Software	HBase 0.94.17	HBase 0.94.17	YCSB
	Hadoop 1.2.1	Hadoop 1.2.1	
	Zookeeper 3.4.5	Zookeeper 3.4.5	
	Mongo 2.4.10	Mongo 2.4.10	
	Cassandra 2.05	Cassandra 2.05	
Functionality	HBase Datanode, Name-node, Secondary Name-node, RegionServer, Zookeeper, Cassandra Daemon, Mongod, Mongo Router, Config Server	HBase Datanode, Cassandra peers, Mongod, Mongo Router (Nodes 2–4)	YCSB client

Due to the fact that the existence of replication indirectly reduces disk space for the original data, the limit of entries to the DB was set to 1.8 Million, given the available resources. Each record consisted of 10 fields and each field of 100 bytes, resulting in 1 KB per record. Thus the overall actual data used in the tested systems were around 5.4 GB.

3.3 Measurement Process and Execution Automation

The actual benchmark execution is performed through the use of YCSB, a client that is responsible for creating queries against the target databases, based on the input parameters that define the type of operations, and for connecting and submitting the queries based on a set of drivers for each system. One key parameter that needs to be clarified is the Throughput (in Ops/sec) which is the desirable number of operations that must be achieved by the system. This does not necessarily mean that the system will achieve this rate however. Latency of the respective operations is also logged and monitored. Average values for these metrics are reported in the end of the measurement cycle. YCSB also contains a set of default workloads that are indicative of specific use cases. Examples of this are Workload A (typical of user session storing for action logging), Workload B (photo tagging in social networks), Workload C (caching of data), Workload D (user status in social networks), Workload E (forum discussions retrieval) and Workload F (user management DBs). More details on these workloads are given in Sect. 4.

The automated framework for the experiment execution appears in Fig. 1. The user inputs a number of parameters such as the number of nodes, the DB size in terms of records, desired throughput, what type of DB to setup etc. This information is then passed to the partial executors, which are responsible for creating the resources on BonFIRE (via OCCI), configuring the nodes via ssh based on the DB type (setting up seed files, alerting where to find the Namenode, creating, populating or cleaning up tables etc.) and finally launching the YCSB client to perform the queries.

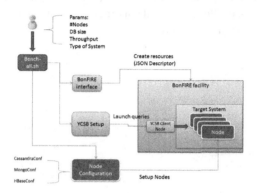

Fig. 1. Automated execution framework

4 Experiments and Results

In order to perform the experiments against the deployed databases, the YCSB benchmarking client was used. YCSB produces queries against the former, based on a throughput that is determined by the parameters of execution. In reality, while this throughput is set, it is perceived as the target limit. However it is limited by the size and endurance of the underlying system. Thus, while the set throughput was starting from 1000 operations per second and increased each time by a thousand, the actual achieved rates were not completely aligned, as will be seen by the measurements. Each measurement (for a given system, workload and set throughput) was performed 4 times and the average was calculated. In many cases there were deviations that can be attributed to the operation of the underlying Cloud service. Timing constraints were also used, meaning that each series of measurements included an overall maximum time for completion. This maximum time was calculated based on the types of actions performed against the DB and the necessary throughput. If that throughput was less than the 2/3 of the needed limit, then the experiment was stopped since it would not add any additional information for the charts and would introduce unnecessary delays. Following, the results per type of workload are presented.

4.1 YCSB Workload a (50 % Updates–50 % Reads)

In workload A, HBase and Cassandra have achieved a significantly higher throughput than MongoDB, as it appears in Fig. 2. The reason for this is the increased updates ratio. MongoDB returns an update success when this is registered in RAM, providing relatively relaxed persistency guarantees, thus this element is not portrayed in the latency. Persistency of data is programmed every 100 ms through the journaling mechanism and the synchronization between the RAM data and disk file data is performed every 60 s. However, due to the limited memory of the used system, it appears that the OS was synchronizing the files in shorter intervals in order to free memory space and fetch the necessary files for the read operations.

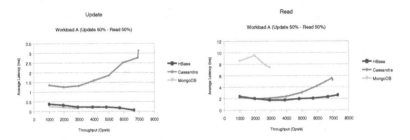

Fig. 2. Comparative Latency vs Throughput in YCSB Workload A

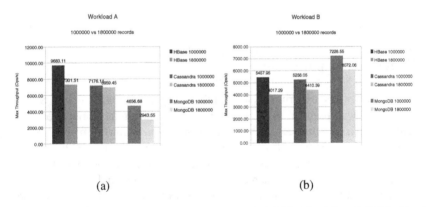

Fig. 3. Throughput Comparison for different DB sizes in (a) Workload A and (b) Workload B (Color figure online)

On the contrary, HBase and Cassandra are not affected by updates since these approaches do not fetch the data to be updated from the disk bit perform the changes in new files (HFiles and SSTables) by using disk serial write capabilities. The update of the data happens in the background by merging these files, without influencing significantly the performance of these systems due to the retrieval times of files from disk. With regard to the effect of data volume in DB performance (Fig. 3), MongoDB seems to be significantly affected (for the same reasons mentioned above) and its performance is reduced by 37 % when operations are increased to 1.8 million, in comparison to

1 million. When data are of a smaller volume, caching mechanisms portray a larger hit ration and databases perform significantly better. HBase also dropped by 25 %, showing the need for more RAM. Cassandra on the other hand was not significantly affected, portraying a deterioration of 3 %.

4.2   YCSB Workload B (5 % Updates–95 % Reads)

In workload B, MongoDB performed significantly better (Fig. 4). The memory mapped files in conjunction with the Zipfian distribution of YCSB, that selects a subset of data to perform the multitude of operations, have enabled the caching mechanisms to be exploited. Lack of increased updates has also helped towards this direction. HBase and Cassandra performed significantly worse than in workload A, due to the limited memory dedication to caching, since this value is a percentage of the Heap size (default 1 GB) used in every node. On the other hand, MongoDB allocated dynamically all the available memory not used by the remaining node operations. This difference is also portrayed in the different DB sizes used, as depicted in Fig. 3b). A second reason for the reduced performance of HBase is the fact that data retrieval is performed from the disk to the volatile memory through the Java Heap process. Thus read intensive workloads cause Heap fragmentation and are managed by the Garbage Collector, putting more strain on the system. HBase in next versions (0.96.3) gives the opportunity to the volatile memory "blockcache", which is the memory part responsible for the caching mechanism, to be decoupled from the Heap size of the RegionServer process, thus exploiting more the available memory [9]. Cassandra in its default setting does not cache data but data keys, making this their retrieval faster. Increased data volumes have caused a performance degradation of 26 % in HBase, 16 % in Cassandra and 19 % in MongoDB (Fig. 3b).

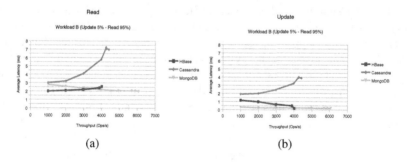

Fig. 4. Comparative Latency vs Throughput in YCSB Workload B

4.3   YCSB Workload C (100 % Read)

In workload C (Fig. 5a), given that it is an exclusively read workload, the issues mentioned in the previous sections are clearly depicted. MongoDB achieves a very low latency due to the lack of updates and synchronization issues. On the contrary, HBase's poor caching strategy under limited available RAM has significantly affected its performance. Of course in cases where the latter does not apply, results could be improved with relation to this case. On the other hand, due to the fact that Cassandra caches the value keys mostly used, it is not affected so intensively by the data read size. However data retrieval on disk makes it less attractive than MongoDB for these kinds of loads. For Cassandra, there is the ability to cache data rows that are requested for reads, however this option was not used since it was not part of the default settings.

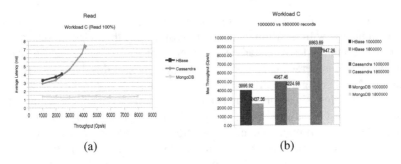

(a)   (b)

Fig. 5. (a) Comparative Latency vs Throughput in YCSB Workload C (b) Throughput Comparison for different DB sizes in Workload C (Color figure online)

As it was anticipated, the increased data volume (Fig. 5b) mainly affected Hbase (39 % degradation). This was significantly lower in the other DBs (17.5 % for Cassandra and 11.5 % for MongoDB).

4.4   YCSB Workload D (95 % Reads–5 % Inserts)

In workload D, a significant difference is the fact that reads are performed with a Latest distribution on the more recently used data and not a Zipfian one. Thus MongoDB again exploits the memory mapped files and portrays a very good performance (Fig. 6). Data insertion does not seem to affect the system's performance, since MongoDB uses mechanisms to preallocate the necessary space. This on the other hand creates an issue of needing too much space for initialization of the DB. Inserts in HBase and Cassandra are performed by creating new files (HFiles and SSTables), thus eliminating the need for concrete data positioning retrieval on the disk during insertion. The usage of Latest distribution helped Cassandra to perform significantly better with relation to workload B and to approach the ratings of MongoDB. This was probably caused by the fact that an increased number of reads were served by the data contained in the system's memT-able. The same behavior was expected from HBase, however the large read number has probably affected also the garbage collection in the JVM. On the contrary, in Cassandra the memory part responsible for storing the keys is off heap. HBase demonstrated a 40 % drop in case of increased data volumes (Fig. 7a), while for Cassandra and HBase the respective percentages were 5 % and 15.1 %.

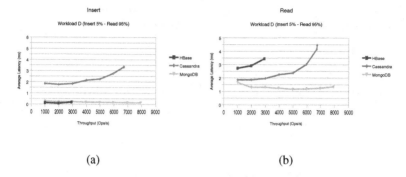

Fig. 6. Comparative Latency vs Throughput in YCSB Workload D

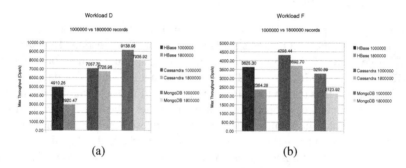

Fig. 7. Throughput Comparison for different DB sizes in (a) Workload D and (b) workload F (Color figure online)

4.5   YCSB Workload F (50 % Reads- 50 % Read/Modify/Write)

In this workload, also caching mechanisms seem to be the key for reaching high performance. Due to the fact that the available RAM in each node was 4 GB (much less than the recommended size), systems that invest in achieving good memory mapping are hindered by the complexity of the workload in this case. Thus MongoDB portrayed a reduced performance (similar to the one in workload (A) while HBase continued to deteriorate as in the cases of B, C and D. On the other hand, Cassandra that does not try to perform these optimizations portrayed less deviation, as it appears in Fig. 8. With relation to DB size (Fig. 7b), HBase had a deterioration in performance by 53 %, the same as MongoDB, while Cassandra was more stable portraying a 16.4 % drop.

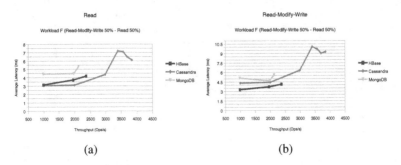

Fig. 8. Comparative Latency vs Throughput in YCSB Workload F

4.6 YCSB Workload E (Scan 95 %–Insert 5 %)

In workload E scan operation, YCSB uses a Zipfian distribution to select a specific datum and then to request the next data in a serial manner from that location. The size of this retrieval is decided via a uniform distribution with a maximum number of 100. In this case HBase and Cassandra performed better than MongoDB since their data are stored serially in files and thus their retrieval is more efficient (Fig. 9). MongoDB on the other hand performed a number of operations per file to retrieve the data from disk, if these data were not located in memory at the time of request. Another reason is that the retrievals in the DB are based on the actual data fields inside the records (due to the document orientation of Mongo) and not based on indexes like in the case of HBase and Cassandra. In terms of database size influence (Fig. 10), Cassandra and HBase seem not to be affected (HBase improves actually its performance by 11 %), however MongoDB suffers from an 88 % drop in performance, due to the limited memory size, type of operation and data access.

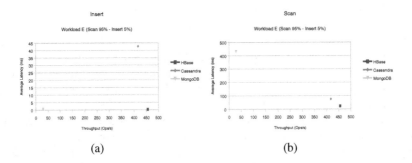

Fig. 9. Comparative Latency vs Throughput in YCSB Workload E

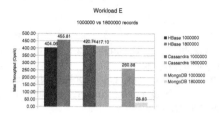

Fig. 10. Throughput Comparison for different DB sizes in Workload E (Color figure online)

5 Conclusions

In conclusion, and following the analysis performed in the previous sections, we can identify cases for which each of the investigated solutions performed optimally. MongoDB was especially efficient in cases where reads constituted the majority of performed operations thanks to the memory-mapped files that it uses. This system however exemplified a particular weakness in scan and retrieval workloads, in which disk operations were necessary. On the other hand, Cassandra's enablement of caching only for the data location on disk resulted in it being the more stable solution despite the change in workload types of the YCSB client and maintained a satisfactory and stable performance in all cases. This was enhanced by the selection of the relaxed consistency option and the usage of a triple replication setup in order to balance the workload in the system. Finally, HBase was severely affected by the choice to store temporary data on-heap especially in conjunction with the limited memory resources in the available nodes. Thus it was not only unable to exploit caching mechanisms, but the latter seemed also to hinder the efficient system operation in some cases.

The systems used in the context of this paper heavily rely on the caching mechanisms enabled in each case and the according design. The experiment aided in identifying several aspects of these mechanisms and their effect on performance, especially under limited available memory in the systems.

During setup and execution, numerous parameters of the examined tools could be toggled in order to adapt to the given tests. This kind of adaptation is worth to consider as future work, extending the insights gained from the initial examination attempted in this work and based on the envisioned usage and deployment by potential adopters of NoSQL technologies and available offerings.

Acknowledgments. This research has been co-financed by the European Union (European Social Fund – ESF) and Greek national funds through the Operational Program "Education and Lifelong Learning" of the National Strategic Reference Framework (NSRF) – Research Funding Program: Thales. Investing in knowledge society through the European Social Fund.

References

1. Digital Universe Infographic.IDC, December 2012. http://www.emc.com/infographics/digital-universe-business-infographic.htm
2. Presto: Interacting with petabytes of data at Facebook. Lydia Chan, November 2013. https://www.facebook.com/notes/facebook-engineering/presto-interacting-with-petabytes-of-data-at-facebook/10151786197628920
3. CERN Computing. http://home.web.cern.ch/about/computing
4. List of NoSQL databases. http://nosql-database.org
5. Han, J., et al.: Survey on NoSQL database. In: 2011 6th International Conference on Pervasive Computing and Applications (ICPCA), 26–28 October 2011, pp. 363–366 (2011). doi: 10.1109/ICPCA.2011.6106531
6. Cooper, B.F., et al.: Benchmarking cloud serving systems with YCSB. In: Proceedings of the 1st ACM Symposium on Cloud computing, pp. 143–154. ACM (2010)
7. Bonfire project Cloud testbeds. http://www.bonfire-project.eu/
8. Open Cloud Computing Interface Standard. http://occi-wg.org/
9. BlockCache 101.Nick Dimiduk. http://www.n10k.com/blog/blockcache-101/. Accessed Sep 2014
10. Poess, M., Floyd, C.: New TPC benchmarks for decision support and web commerce. ACM SIGMOD Rec. **29**(4), 64–71 (2000)
11. Shi, Y., et al.: Benchmarking cloud-based data management systems. In: Proceedings of the Second International Workshop on Cloud Data Management (CloudDB 2010) (2010)
12. Hecht, R., Jablonski, S.: NoSQL evaluation: a use case oriented survey. In: Proceedings of the 2011 International Conference Cloud Service Computing, pp. 336–341 (2011)
13. Rahman, M.R., et al.: Toward a principled framework for benchmarking consistency. In: Proceedings of the Eighth USENIX Conference on Hot Topics in System Dependability. USENIX Association (2012)
14. Golab, W., et al: Analyzing consistency properties for fun and profit. In: PODC 2011: Proceedings of the 30th Annual ACM SIGACT-SIGOPS Symposium on Principles of Distributed Computing, pp. 197–206 (2011)
15. Van der Veen, J.S., et al.: Sensor data storage performance: SQL or NoSQL, physical or virtual. In: 2012 IEEE 5th International Conference on Cloud Computing (CLOUD). IEEE (2012)
16. Abadi, Daniel J.: Data management in the cloud: limitations and opportunities. IEEE Data Eng. Bull. **32**(1), 3–12 (2009)
17. Iosup, A., et al: On the performance variability of production cloud services. In: 2011 11th IEEE/ACM International Symposium on Cluster, Cloud and Grid Computing (CCGrid). IEEE (2011)
18. Brewer, E.A.: Towards robust distributed systems. In: Symposium on Principles of Distributed Computing (2000)
19. Sakr, S., et al.: A survey of large scale data management approaches in cloud environments. IEEE Commun. Surv. Tutorials **13**(3), 311–336 (2011)

Cloud Elasticity: A Survey

Athanasios Naskos[1]([✉]), Anastasios Gounaris[1], and Spyros Sioutas[2]

[1] Department of Informatics, Aristotle University of Thessaloniki,
Thessaloniki, Greece
{anaskos,gounaria}@csd.auth.gr
[2] Department of Informatics, Ionian University, Corfu, Greece
sioutas@ionio.gr

Abstract. Cloud elasticity is a unique feature of cloud environments, which allows for the on demand (de-)provisioning or reconfiguration of the resources of cloud deployments. The efficient handling of cloud elasticity is a challenge that attracts the interest of the research community. This work constitutes a survey of research efforts towards this direction. The main contribution of this work is an up-to-date review of the latest elasticity handling approaches and a detailed classification scheme, focusing on the elasticity decision making techniques. Finally, we discuss various research challenges and directions of further research, regarding all phases of cloud elasticity, which can be deemed as a special case of autonomic behavior of computing systems (This research has been co-financed by the European Union (European Social Fund - ESF) and Greek national funds through the Operational Program "Education and Lifelong Learning of the National Strategic Reference Framework (NSRF) - Research Funding Program: Thales. Investing in knowledge society through the European Social Fund.").

1 Introduction

Cloud computing forms a deployment model, which aims to reduce the momentary cost of the computing resources through the leasing of dynamically adjusted virtual resources, which can be occupied on-demand. Virtual resources are virtual versions of actual resources, most commonly in the form of Virtual Machines (VMs), which leverage the virtualization technology [65]. The offered pay-as-you-go pricing model accompanied by the elastic resource handling, has assisted the wide adoption of the cloud deployments, as the client is obliged to pay only for the used resource. As such, cloud computing has managed to make the provision of remote computing resources (e.g., VMs) the main option not only for scientific institutions but any size of organizations and enterprises. However, the efficient resource handling is a key aspect to the deployment cost reduction.

There are numerous works that propose various cloud elasticity handling mechanisms. In this work, our focus is on all aspects of elasticity, but we particularly aim to shed light on the decision making mechanisms in relation with the underlying models employed. Additionally, through our taxonomy, we aim to

© Springer International Publishing Switzerland 2016
I. Karydis et al. (Eds.): ALGOCLOUD 2015, LNCS 9511, pp. 151–167, 2016.
DOI: 10.1007/978-3-319-29919-8_12

render the various techniques, which nowadays tend to be developed in isolation, more comparable with each other.

We regard elasticity techniques as an interdisciplinary field of two main areas: distributed/cloud computing and autonomic computing. As a field of autonomic computing, it comprises all four phases of the MAPE loop [44], namely *Monitoring, Analysis, Planning* and *Execution*. Each distinct phase presents unique research challenges, which are addressed by the presented works with various approaches. In this work, we mostly focus on the last three phases.

Some efforts to create an overview of the cloud elasticity area have been made in the past, for example [26] is complementary to our work, but it focuses more on the tools, the benchmarks and the workloads. We present elasticity strategies in a more broader fashion as we elaborate more on the elasticity decision mechanism. [48] is also complementary to our work, but we present more up-to-date proposals and cover a more extended range of elasticity actions and objectives. An older and narrower survey has also appeared in [33]. A general systematic review about commercial cloud services is conducted in [46], where the authors present the main challenges regarding the elasticity property. As such, our work fills an important gap on a timely issue.

The structure of this survey paper is as follows. In Sect. 2, we present the taxonomy and the classification table. In Sect. 3, we delve into more details for each classification dimension of our taxonomy and we outline the main findings. We conclude in Sect. 4.

2   Taxonomy and Classification

In order to provide a concise classification of the existing approaches to cloud elasticity, we first propose a taxonomy that will enable our work to shed light on the differentiating aspects of the various proposals. The taxonomy is summarized in Fig. 1 and consists of the following dimensions:

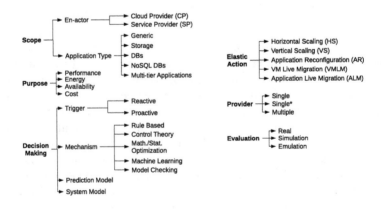

Fig. 1. Classification scheme

- *Scope.* This aspect is divided into two classification categories (i) the *Enactor* and the (ii) *Application Type*. The former indicates whether the elastic technique is applied by the cloud infrastructure provider (*Cloud Provider (CP)*) or the user of the cloud infrastructure, who deploys and manages cloud applications on top of the cloud infrastructure (*Service Provider (SP)*). *Application Type* indicates whether the proposal refers to the elastic handling of a particular type of cloud application from the following list: relational databases (*DBs*), NoSQL databases (*NoSQL DBs*), *Multi-tier Applications* (e.g., typical business web applications), *Generic* (if the tool is application-agnostic) or *Storage*.
- *Purpose.* In this dimension, we classify the techniques according to the purpose of elasticity actions. The purpose can be one of the following: (i) *Performance*, (ii) *Availability*, (iii) *Cost*, (iv) *Energy*. *Performance*, refers to the maintenance or guarantee of acceptable, either user or Service Level Agreement (SLA) specified, application performance. *Availability* refers to the degree to which applications and resources are in an operable and committable state at the time point when they are needed by end users [42]. *Cost* refers either to the reduction of the operational cost of the application deployed in the cloud, commonly also maintaining the *Performance* goal, or to the maintenance of cost thresholds under specific performance constraints. Finally, the *Energy* category, is closely related to the *Cost* one but covers elastic techniques, which directly aim at minimizing the energy footprint.
- *Decision Making.* There are four distinct categorization criteria that characterize the decision making procedure of every work in our taxonomy, namely (i) *Trigger*, which indicates whether the elasticity mechanism is triggered in a reactive or proactive manner; (ii) *Mechanism*, which refers to the decision making methodology; (iii) *Prediction Model (PM)*, which denotes the utilization of a model to predict future incoming load variations or specific measurement values; and (iv) *System Model (SM)*, which refers to the utilization of a model to represent the (elastic) behavior of the system, on top of which the complete elasticity policy is built (e.g., queues). Elasticity mechanisms are further classified into the following categories: (1) *Rule Based*, (2) *Mathematical/Statistical Optimization*, (3) *Machine Learning*, (4) *Control Theory* and (5) *Model Checking* according to the main field to which the elasticity policy belongs.
- *Elastic Action.* Cloud resource elasticity may be applied in different forms and can refer to modifications in (i) the size (*Vertical Scaling (VS)*), (ii) the location (*VM Live Migration (VMLM)*) or (iii) the number of VMs employed (*Horizontal Scaling (HS)*). Examples of these three elasticity types are the allocation of more memory or CPU to a VM, moving a VM to a less loaded physical machine and increasing the number of VMs of an application cluster, respectively. *Elastic Action* additionally includes two other elasticity types, (iv) the *Application Reconfiguration (AR)*, where the elastic tool is capable of handling specific application aspects (e.g., DB cache size) and (v) *Application Live Migration (ALM)*, where only application-specific components are migrated instead of the full VM, such as database instances.

- *Provider.* This classification category refers to the number of cloud infrastructure providers that the elastic tool supports simultaneously. The possible values are (i) *Single*, which denote that only one cloud provider is supported, (ii) *Single**, which denotes that more than one providers are supported, however not simultaneously and (iii) multiple, where the elasticity control is spread across multiple cloud providers simultaneously.
- *Evaluation.* Finally, the last aspect refers to the type of the *Evaluation* of every work. The possible values are: (i) *Simulation*, where the results are obtained based on computations on a simulated artificial environment (e.g., OMNeT++), (ii) *Emulation* where the evaluations results are obtained in an artificial environment that behaves according to real-world traces, and (iii) *Real*, where the elastic tool is applied on a real cloud infrastructure.

Based on the taxonomy above, we classify the existing proposals for cloud elasticity as shown in Table 1. The taxonomy above does not cover the type of the feedback information collected by the environment to drive the elasticity decision making and enforcement, because the type of the feedback seems to play a less important role in classifying the proposals. More specifically, all proposals utilize a mechanism to monitor specific system/network/application-specific metrics to assist the decision making. To deal with possible load spikes or measurement instabilities, many works utilize smoothing techniques like Exponential Weighted Moving Average (EWMA), Exponential Moving Average (EMA) or just Moving Average (MA). Further details are omitted due to space constraints.

3   Overview of Existing Solutions

In this section, we provide details with respect to the main solution approaches for each taxonomy dimension in turn.

3.1   Scope

The first aspect of the scope dimension indicates who is responsible for the elasticity mechanism. In several proposals, the elasticity technique is bundled with the core cloud infrastructure and the corresponding techniques are described as *Cloud Provider*-specific. For example, [37] relies on a tool that is installed on top of IaaS infrastructures, and the DejaVu system in [68] extends the functionality of such infrastructures. Another set of proposals require special privileges to resources (e.g., [54] depends on a custom KVM module and interface, [21] is integrated into OpenStack, and [61] configures the CPU voltage and frequency) or access to information that only a cloud provider is able to provide (e.g., [50] depends on physical machine local information). Nevertheless, the majority of proposals enable the provision of advanced elasticity for cloud-based services regardless of or extending the built-in elasticity functionalities of the cloud providers; those are referred to as *Service Provider*-specific.

Regarding the application type on which the proposals focus, most of them are either application independent or tailored to web service-based multi-tier

Table 1. The classification of research proposal according to out taxonomy.

Citation	Scope		Purpose	Decision making				Elastic action	Provider[a]	Evaluation
	Enactor	Appl. type		Trigger	Mechanism	PM	SM			
[54]	CP	Generic	Perf.	Proactive	Math./stat. optimization	×	×	HS VMLM	Single	Real
[21]	CP	Generic	Perf.	Reactive	Rule based			HS	Single[a]	Real
[37]	CP	Generic	Perf.	Reactive	Rule based			HS VS	Single	Real
[50]	CP	Generic	Perf. energy	Reactive	Math./stat. optimization			VMLM	Single	Real
[61]	CP	Generic	Perf. energy	Reactive proactive	Rule based math./stat. optimization	×		VS VMLM	Single	Real
[68]	CP	Multi-tier applications	Perf.	Proactive	Mach. learn.		×	HS VS	Single	Real
[31]	SP	DBs	Perf.	Reactive	Mach. learn.		×	ALM	Single	Real
[20]	SP	DBs	Perf.	Reactive proactive	Rule based	×		HS VMLM ALM	Single	Real
[59]	SP	DBs	Perf.	Proactive	Rule based math./stat. optimization			HS ALM	Single	Real
[55]	SP	Generic	Avail.	Reactive	Rule based			HS	Multiple	Real
[63]	SP	Generic	Perf.	Proactive	Mach. learn.	×	×	VS VMLM	Single	Real
[39]	SP	Generic	Perf.	Proactive	Rule based math./stat. optimization	×	×	HS	Single	Real
[60]	SP	Generic	Perf.	Reactive	Control theory	×	×	HS	Single[a]	Real
[49]	SP	Generic	Perf.	Reactive	Rule based		×	HS	Single[a]	Real
[17]	SP	Generic	Perf.	Reactive proactive	Rule based	×		HS	Single	Simulation
[16]	SP	Generic	Perf.	Reactive proactive	Control theory		×	HS	Single	Simulation
[51,52]	SP	Generic	Perf.	Reactive proactive	Rule based mach. learn.	×	×	HS	Single	Simulation
[29]	SP	Generic	Perf. cost	Proactive	Mach. learn. math./stat. optimization	×	×	HS VS	Single	Real
[25]	SP	Generic	Perf. cost	Proactive	Rule based		×	HS AR	Multiple	Real
[66]	SP	Generic	Perf. cost	Reactive	Rule based		×	HS VS AR	Single	Real
[40]	SP	Multi-tier applications	Perf.	Proactive	Control theory			VS	Single	Real

(*Continued*)

Table 1. (Continued)

Citation	Scope	Appl. type	Purpose	Decision making		PM	SM	Elastic action	Provider[a]	Evaluation
	Enactor			Trigger	Mechanism					
[43]	SP	Multi-tier applications	Perf.	Reactive proactive	Rule based math./stat. optimization		x	HS	Single	Real
[38]	SP	Multi-tier applications	Perf.	Reactive proactive	Rule based math./stat. optimization		x	HS	Single	Real
[35]	SP	Multi-tier applications	Perf. avail.	Reactive	Rule based		x	HS	Single	Real
[56]	SP	Multi-tier applications	Perf. avail.	Reactive	Rule based			HS VS	Multiple	Real
[32]	SP	Multi-tier applications	Perf. cost	Proactive	Math./stat. optimization	x	x	HS	Single	Real
[22]	SP	Multi-tier applications	Perf. cost	Proactive	Math./stat. optimization		x	HS AR	Single	Simulation
[18]	SP	Multi-tier applications	Perf. cost	Proactive	Rule based control theory	x		HS AR	Single	Real
[36]	SP	Multi-tier applications	Perf. cost	Reactive	Rule based math./stat. optimization		x	HS	Single	Real
[57]	SP	Multi-tier applications	Perf. cost	Reactive	Rule based math./stat. optimization			HS	Single	Emulation
[62]	SP	Multi-tier applications	Perf. cost	Reactive proactive	Rule based	x		VS	Single	Real
[41,45,64]	SP	NoSQL DBs	Perf.	Proactive	Mach. learn.		x	HS	Single[a]	Real
[58]	SP	NoSQL DBs	Perf.	Proactive	Mach. learn. math./stat. optimization		x	HS AR	Single	Real
[53]	SP	NoSQL DBs	Perf.	Proactive	Model checking mach. learn. math./stat. optimization		x	HS	Single[a]	Emulation
[15]	SP	NoSQL DBs	Perf.	Reactive	Control theory mach. learn.		x	HS	Single	Real
[23]	SP	NoSQL DBs	Perf.	Reactive	Rule based			HS AR	Single	Real
[47]	SP	NoSQL DBs	Perf.	Reactive	Rule based control theory			HS	Single	Real
[27]	SP	NoSQL DBs	Perf.	Reactive	Rule based math./stat. optimization			HS AR	Single	Real
[19]	SP	Storage	Perf.	Reactive proactive	Rule based			HS	Single	Real

[a]Multiple providers supported, not simultaneously

applications. Most of the latter proposals support the elastic handling of all three tiers of a web application (i.e., Web Server, Application Server, Storage Server tiers), except from [18,32], which handle only the elasticity of the Application Server tier, and [62,68], which simply target web services. A significant portion of elasticity proposals targets the NoSQL area. The techniques in this group are either system-specific (e.g., [23] targets Cassandra, [27] targets HBase, [58] targets Infinispan, while [47] considers HDFS) or applicable to a larger set of NoSQL systems, such as Cassandra, HBase, Voldemort and Infinispan [15,41, 45,53,64]. Elasticity in relational databases is considered by [20,31,59]. [20,31] can be used with any relational database as they do not modify the database engine, while in [59], the database engine is modified to support live migrations employing a technique inspired by [30]. Finally, there is a single proposal that is categorized as *Storage* [19], where the elasticity handling of storage functionality in virtual machines is considered, through caching techniques.

3.2   Purpose

All the techniques appearing in Table 1 aim to improve performance. The only exception is [55], where the elasticity aim is the increased availability through the utilization of multiple cloud providers. The performance goal can be either fixed (e.g., in SLAs or stated as user-defined thresholds) or expressed as continuous monitoring and optimization of the system utility. In such works, the monetary cost reduction is typically indirectly considered, through the pursue of utilizing as few resources as possible while maintaining acceptable performance. In [35,56], the performance goal is coupled with offering *Availability* guarantees.

Several proposals target financial cost reduction explicitly. More specifically, [29,36,57] employ cost estimation to scale-in or -out the cloud resources, while [25] use similar estimates to select between deployment on public or private cloud infrastructure, and [62] performs a Return on Investment (ROI) analysis before the actual deployment. Other techniques handle elasticity according to budget limits. More specifically, [66] prevents scaling if the maximum available cost is exceeded, [22] enforces an application reconfiguration (i.e. textual server responses for bandwidth saving) to keep the cost below the budget limit and [32] offers a budget classification (i.e. metal classification: gold/silver/bronze), which configures the resource scaling limits. Finally, [18] tries to co-locate several applications on the same VM to reduce the provisioning cost.

Additionally, there are two works that consider the *Energy* saving combined with the *Performance* purpose. In [50] live migration is employed to set as many machines to sleep mode as possible, whereas, in [61], VM resources are subject to dynamic voltage and frequency scaling to save energy.

3.3   Decision Making

Triggering of Decision Making Process. The works are divided into (i) *Reactive*, (ii) *Proactive* and (iii) combined *Reactive and Proactive*. On the one hand, *Reactive* approaches are typically based on the continuous monitoring of specific metrics and the validation of threshold-based rules. Most commonly, upon a single threshold violation, the decision making process is triggered. However, the decision process can be also triggered only after a pre-specified time period of threshold violations (e.g., [55,57]), or a pre-specified number of violating measurements (e.g., [21]), to avoid over-reacting. On the other hand, *Proactive* approaches employ a mechanism to predict the future load variation and/or the future behavior of the system. However, purely proactive approaches tend to suffer from the fact that they are not able to cope with sudden workload spikes. To overcome this concern, works like [17,20,43] adopt a hybrid approach. In addition, [16,38] propose the utilization of reactive approaches for scaling-out and proactive approaches for scaling-in. The former is used to allow for quick adaptations to workload spikes. There are also proposals that utilize reactive approaches when the proactive mechanism is uncertain about the decision [61], or when the predictor is not adequately trained to take a proper decision [51,52]. [62] proposes the combination of reactive and proactive techniques, where the latter is activated on a daily basis. Finally, in [19], the reactive and proactive approaches are not concurrently activated, but the system can support any of them separately.

Decision Making Mechanism. In the previous section, we mentioned the main methodologies used for elasticity decision making. Here, we elaborate on this issue, describing the application first of single methodologies and then of hybrid solutions.

However, the corresponding techniques need not be simple. For example, [17] utilizes a bunch of concurrent prediction models to estimate load and check for potential future threshold violations. Also, [20] uses a prediction model to estimate the load for proactive scaling based on specified rules. System modeling in general enhances the decision policies. In [49], the system is modeled as a queue of jobs and an elasticity action is triggered upon a job arrival or completion. Two interesting rule-based approaches that build on non-trivial system models are the [35], where the system is modeled as an automaton moving to different states because of rule enforcement, and [25], where a graph model that captures the impact of elasticity rules on the entire system is adopted. As a final example, in [39], a fuzzy rule-based approach is followed, where the user specifies rules in the form *"IF the workload is high, AND response-time is slow, THEN add two more VMs to the existing resources"*, without the need of characterizing the "high" and "slow" values; those values are computed based on information provided by technical stakeholders.

Mathematical/Statistical Optimization-Based Policies. These techniques model the elasticity problem as an optimization one. In [32], the optimal scaling strategy is found following a branch-and-bound technique after having performed sophisticated time-series analysis to predict future external load. In [22], elasticity decision making is reduced to a utility maximization problem amenable to dynamic programming; this technique employs a queue model and model checking as a pre-processing step to quantify the potential benefits of the employment of different types of algorithms for self-adaptation. The approach in [50] leverages Bernoulli trials to find appropriate VM placing to guide the live migration. Finally, optimization may refer to system modeling itself that then drives elasticity; for example, [54] uses online profiling and curve fitting to yield a performance model, which can predict whether the application is going to violate a target.

Machine Learning-Based Policies. Machine learning is commonly used in elasticity decision making. [68] builds a system model in the form of a classifier, while also clusters workloads in representative groups. Past elasticity decisions for the same group influence future decisions. [31]'s approach is similar. In [63], a markov-chain-based prediction model provides estimates that are fed to a multi-variant classifier in order to classify future states as either normal or anomalous, and take elasticity actions accordingly. An example of more advanced techniques is in [41,45,64], where a Q-Learning approach is followed to compute the optimal action-state values in order to indirectly solve a Markov Decision Model (MDP) describing the system.

Control Theoretical Policies. Control theory, being a scientific field capable of providing principled autonomic computing solutions, has been adopted by certain elasticity policies. As an example, [60] follows a control theoretical approach that builds on top of a queue modeling representation of the system and also employs a predictor. Also, in [40], an example of using Kalman filters and feedback controllers to drive elasticity is provided. Finally, [16] discusses a controller scheme that combines proactive and reactive policies. As in other similar works, the cloud infrastructure is modeled as a queue, while estimators for future external load are assumed to be in place.

Hybrid Policies. The afore-mentioned policies correspond to decision modules that employ one of the specified mechanisms. However, elasticity techniques often employ several such mechanisms, as described below.

The most common hybrid approach is to combine rules with one of the rest mechanisms. Rules can effectively extend control theoretical solutions. For example, in [47], an integral controller is proposed, which is based on proportional thresholding to dynamically adjust the upper and lower CPU utilization thresholds used for elasticity decisions. In [18], linear regression is used to predict the future load, and subsequently, the predicted values are fed into a custom model-free proportional-derivative controller. The final decisions about the number of VMs to be added or removed are taken according to a rule-based policy. Rules can be combined with machine learning techniques as well. A representative of this hybrid category is in [51,52], where three models from the WEKA tool are used to support the decision making. The first model is a time series forecaster that estimates the future workload. The second model is an incrementally updateable Naive Bayes model that learns the relationship between the current workload and a classification schema of threshold violation, and the third model is also an updateable Naive Bayes model which estimates the optimal number of VMs.

Another family of hybrid solutions combine rules with some form of optimization. In [43], the system is modeled as a closed form queueing network, where mean value analysis is used to compute the queue lengths, and the response time, throughput and utilization of the system. Following an iterative optimization technique based on Binary Search Trees, the technique tries to minimize the number of VMs needed at each tier without violating performance thresholds. In [57], first an optimization problem that finds the number of VMs maximizing the application profit is solved, before dynamically generating the elasticity rules. [27] first evaluates a rule to determine if a scaling action is required, and if this is the case, a variant of a bin-packing problem is solved. [36] also uses rules to detect workload changes, and then runs an algorithm to decide on VM additions and removals at each layer of a multi-tier application, so that the response time of the application is below a specified threshold and the deployment cost is minimized. Another form of combination of mechanisms appears in [38], where a rule-based reactive technique is used to scale out the resources, while a more elaborate predictive technique, based on regression models (System Model), is used to scale in. [61] is based on rules and prediction models and an interesting feature is that it directly tackles prediction error through an adaptive padding technique.

The last family of hybrid techniques are those that combine machine learning with either optimization or control theory. In [29], the resource requirements are continuously estimated according to the expected workload. The workload is predicted using a polynomial approximation technique and then classified to a set of workload classes. Then, a two-phase technique runs. First, the optimal VM size (i.e., CPU and RAM amount) and the corresponding throughput is specified, thus defining the potential need for vertical scaling. In the second phase, the optimal number of VMs of the specified size is computed, thus defining the potential need for horizontal scaling. In [58], neural networks are used to estimate the throughput and response time of the system and then, a controller solves a constraint optimization problem to determine an optimal resource configuration in terms of number of VMs and data replication degree. In [15], a feedforward controller is used, which monitors the workload and uses a logistic regression model to predict whether the workload will cause SLA violations and react accordingly. This controller is combined with a feedback controller, which monitors the performance and reacts based on the amount of deviation from the desired performance specified in the Service Level Objective (SLO). Finally, in [53], the system behavior for a given external load is clustered in representative groups. This helps to instantiate descriptive models of horizontal scaling in the form of Markov Decision Processes, which are optimally solved. A unique feature of this work is that it employs in parallel model checking to provide probabilistic guarantees regarding the expected performance of elasticity actions.

3.4   Elastic Action

The big majority of works on elasticity considers only *Horizontal Scaling*, where the number of VMs is modified on the fly, e.g., [18, 22, 45, 53, 56, 64]. There are also works that utilize only *Vertical Scaling*, e.g., dynamically configuring the

CPU [40] and the RAM and Disk size [62]. We also report a single technique that performs VM *Live Migration* [50] and a work that performs *Application Live Migration* [31], where only specific databases are migrated, instead of full VMs. However, there are techniques that combine multiple actions. *Horizontal Scaling* along with *Application Reconfiguration* is considered in [23,27,58]. The applied reconfiguration varies between the proposals. More specifically, in [58], the replication degree is configured dynamically. In [23], the cache size is dynamically controlled, while [27] can scale the maximum number of data partitions per node. [29,37,68] combine *Horizontal and Vertical Scaling*. From the techniues that use a fixed number of VMs, [61,63] combine *Vertical Scaling* and *Live Migration*. *Horizontal Scaling* is also combined with *Live Migration* [54] and *Application Live Migration* [59]. Finally, there are two works that combine three types of resource elasticity. [20] uses *Horizontal Scaling* and *DB and VM Live Migration*, while [66] uses *Horizontal Scaling*, *Vertical Scaling* (i.e. CPU and RAM configuration) and *Application Reconfiguration* (i.e., application architectural changes).

3.5 Provider

Most of the works support a *Single* cloud provider, either public or private. Some works support more than one provider, however not simultaneously. More specifically, all of these works are compatible with Amazon-EC2 and also support Grid5000 (used by [60]), Nimbus-based cloud platform (used by [49]), OpenNebula (used by [66]), Eucalyptus (used by [38]), DAS-4 (i.e. a multi-cluster system hosted by universities in the Netherlands used by [32]) and OpenStack (used by [41,45,64]). Finally, there are works that handle the elasticity between multiple cloud providers simultaneously, such as [55,56] that are deployed on a variety of private and public cloud infrastructures.

3.6 Evaluation

As presented in Table 1, most of the works use *Real* deployments for the evaluation of their proposals. The RUBiS benchmark [11] is used in [35,38,40,54,68], the TCP-[C/W] [13,36,37,60]. Another popual benchmark is YCSB [24] used in [15,20,23,41,45,64]. [60] additionally utilizes the Apache Hadoop [1] with the MRBS benchmark suite [6]. Some works use both TPC and YCSB [27,59] or both RUBiS and IBM System S [34,61,63]. CloudStone [3] is used in [29,47]; the latter uses the Olio web 2.0 toolkit [9] of the CloudStone in combination with the Faban workload driver [4]. MediaWiki [7] with the WikiBench benchmark [14] is used in [32] and the FIO micro-benchmarking tool [5] is used in [19] in combination with the USR-1 trace of the MSR traces [8]. The Apache JMeter [2] is used in [39]. Finally, there are other proposals that utilize custom made benchmarks like [21,31,43,58].

There are also works that perform evaluation using simulations [16,17,22,51,52], e.g., utilizing the R statistics suite [10], or OMNeT++ [67]. The simulations can also encapsulate benchmarks, such as the one in [12]. Finally, [53,57] use emulations. [57] uses the FIFA 1998 World Cup traces and the Amazon EC2 payment policy, while [53] uses real traces from a Cassandra NoSQL cluster.

3.7   Discussion and Research Challenges

In this survey we examined the *Analysis*, *Planning* and *Execution* aspects of the state-of-the-art in cloud elasticity. In the *Analysis* phase, the retrieved measurements are used to examine the current state of the system (e.g., whether it is under- or over-utilized) and/or estimate the future load variations. In the former case, a practical approach to determine the current state of the system is to use threshold-based rules. However, the specification of such rules is not a simple task, as it depends on the application needs and demands system administrative skills. To overcome this concern, various approaches are proposed, like the fuzzy rule specification, where the stakeholders knowledge is already analyzed and stored, allowing the user to define high-level thresholds, which are automatically mapped to concrete threshold-based rules. Other approaches propose the transformation of the SLAs and SLOs to threshold based rules, utilizing custom SLA and SLO specification languages and rule conflict solving mechanisms. In the case of future load prediction, there are numerous approaches, which deal with the prediction inaccuracies. There are works that attempt to bound the prediction error, or take inaccuracies into consideration. There are also proposals that utilize more than one prediction algorithms implementing mechanisms to select the most appropriate one based on the current workload.

Knowing the current state of the system and/or the future load variations, in the *Planning* phase, the actual elastic decision should be made. However, this step also hides some difficulties like the decision between the scale-in or scale-out, the selection of the appropriate elasticity type or the determination of the degree of scaling. To deal with these decisions, various approaches are used like pre-specified rules (i.e. the simplest form of planning), utility functions, system models, prediction mechanisms, machine learning techniques or any combination of the previous, to obtain the system behavior before the actual decision enforcement. Each approach has its own drawbacks, as discussed below:

- The usage of pre-specified rules restricts the flexibility of the application, as the amount and the type of scaling is defined a-priory. To deal with this concern, dynamic rule specification or rule update has been proposed in the literature.
- The optimization of a utility function, which contains appropriately selected and weighted metrics, can lead to an acceptable trade-off between contradicting scaling options, however the specification of such a function demands special administrative knowledge. To overcome these concerns, fixed utility functions are proposed, which are generic enough to be applied to many systems.
- The specification of a system model is not a trivial task as it is difficult to create a reliable model that maps the input and output variables of the system. In addition, the system model hinders the flexibility of the elastic mechanism

as, after any structural change of the system, the model needs to be rebuilt. To this end, model generators are proposed, which generate a model without user interference.

– The approaches that utilize system behavior predictions suffer from prediction inaccuracies. The proposed solution is similar to those mentioned for the analysis phase.

– A promising approach is the utilization of machine learning, where the elastic mechanism is trained before its actual deployment. The training phase can also be applied during the actual deployment, allowing for dynamical training. However, in the latter case, the mechanism may not able to handle the elasticity well from the beginning of the deployment. To avoid wrong decisions that under- or over-provision the system, proposals tend to use a threshold-based reactive mechanism until the mechanism is considered well-trained and capable of efficiently handling the elasticity. An interesting related discussion is also in [28].

In the *Execution* phase, the actual elastic decision is enforced through the elastic manager orchestration. The elastic manager is either a standard mechanism provided by the cloud providers as a service or through a remote API (e.g., Amazon EC2 Auto Scale service), or an external manager.

Aspects Requiring Further Research. As a final observation, although a big set of elasticity proposals exists and a considerable amount of them deal with multiple objectives, no systematic solution for dynamic multi-objective optimization under several conflicting objectives, e.g., guaranteeing Pareto optimality, has been proposed. We believe that this is an interesting direction for future work. Another interesting direction is to provide frameworks that can combine several of the solutions that are now isolated. Finally, more research on benchmarks is needed to better assess the quality of each of the proposals.

4   Summary

This survey aims to classify and provide a concise summary of the several proposals for cloud resource elasticity today. We presented a taxonomy covering a wide range of aspects, and we discussed details for each of the aspects, and the main research challenges. Finally, we proposed fields that require further research.

References

1. Apache hadoop. https://hadoop.apache.org/. Accessed 11 Jun 2015
2. Apache jmeter: Graphical server performance testing tool. http://jmeter.apache.org/. Accessed 11 Jun 2015
3. Cloudstone. http://parsa.epfl.ch/cloudsuite/web.html. Accessed 11 Jun 2015

4. Faban: Performance workload creation and execution framework. http://faban.org/. Accessed 11 Jun 2015
5. Fio: A micro-benchmarking tool. http://freshmeat.net/projects/fio. Accessed 11 Jun 2015
6. Hadoop mapreduce dependability, performance benchmarking. http://sardes.inrialpes.fr/research/mrbs/. Accessed 11 Jun 2015
7. Mediawiki: Web hosting benchmark. http://www.wikibench.eu. Accessed 11 Jun 2015
8. Msr cambridge traces. http://iotta.snia.org/traces/388. Accessed 11 Jun 2015
9. Olio web 2.0 toolkit. http://incubator.apache.org/projects/olio.html. Accessed 11 Jun 2015
10. The r project for statistical computing. http://www.r-project.org. Accessed 11 Jun 2015
11. Rubis: Rice university bidding system. http://rubis.ow2.org. Accessed 11 Jun 2015
12. Specjenterprise benchmark system. https://www.spec.org/jEnterprise2010/. Accessed 11 Jun 2015
13. Tcp. http://www.tpc.org. Accessed 11 Jun 2015
14. Wikibench: Web hosting benchmark. http://www.wikibench.eu. Accessed 11 Jun 2015
15. Al-Shishtawy, A., Vlassov, V.: Elastman: elasticity manager for elastic key-value stores in the cloud. In: ACM Cloud and Autonomic Computing Conference, CAC 2013, Miami, FL, USA, 5–9 August 2013, p. 7 (2013)
16. Ali-Eldin, A., Tordsson, J., Elmroth, E.: An adaptive hybrid elasticity controller for cloud infrastructures. In: 2012 IEEE Network Operations and Management Symposium (NOMS), pp. 204–212 (2012)
17. Almeida Morais, F., Vilar Brasileiro, F., Vigolvino Lopes, R., Araujo Santos, R., Satterfield, W., Rosa, L.: Autoflex: service agnostic auto-scaling framework for IaaS deployment models. In: 2013 13th IEEE/ACM International Symposium on Cluster, Cloud and Grid Computing (CCGrid), pp. 42–49 (2013)
18. Ashraf, A., Byholm, B., Porres, I.: Cramp: cost-efficient resource allocation for multiple web applications with proactive scaling. In: 2012 IEEE 4th International Conference on Cloud Computing Technology and Science (CloudCom), pp. 581–586 (2012)
19. Bairavasundaram, L.N., Soundararajan, G., Mathur, V., Voruganti, K., Srinivasan, K.: Responding rapidly to service level violations using virtual appliances. SIGOPS Oper. Syst. Rev. **46**(3), 32–40 (2012)
20. Barker, S.K., Chi, Y., Hacigümüs, H., Shenoy, P.J., Cecchet, E.: Shuttledb: database-aware elasticity in the cloud. In: 11th International Conference on Autonomic Computing, ICAC 2014, Philadelphia, PA, USA, 18–20 June 2014, pp. 33–43 (2014)
21. Beernaert, L., Matos, M., Vilaça, R., Oliveira, R.: Automatic elasticity in openstack. In: Proceedings of the Workshop on Secure and Dependable Middleware for Cloud Monitoring and Management, p. 2 (2012)
22. Cámara, J., Moreno, G.A., Garlan, D.: Stochastic game analysis and latency awareness for proactive self-adaptation. In: SEAMS, pp. 155–164 (2014)
23. Chalkiadaki, M., Magoutis, K.: Managing service performance in the cassandra distributed storage system. In: IEEE 5th International Conference on Cloud Computing Technology and Science, CloudCom 2013, Bristol, UK, 2–5 December 2013, vol. 1, pp. 64–71 (2013)

24. Cooper, B.F., Silberstein, A., Tam, E., Ramakrishnan, R., Sears, R.: Benchmarking cloud serving systems with YCSB. In: Proceedings of the 1st ACM Symposium on Cloud Computing, pp. 143–154 (2010)
25. Copil, G., Moldovan, D., Truong, H.L., Dustdar, S.: On controlling cloud services elasticity in heterogeneous clouds. In: 2014 IEEE/ACM 7th International Conference on Utility and Cloud Computing (UCC), pp. 573–578 (2014)
26. Coutinho, E.F., de Carvalho Sousa, F.R., Rego, P.A.L., Gomes, D.G., de Souza, J.N.: Elasticity in cloud computing: a survey. Ann. Telecommun.-Annales des TéléCommuni **70**, 289–309 (2015)
27. Cruz, F., Maia, F., Matos, M., Oliveira, R., Paulo, J., Pereira, J., Vilaça, R.: Met: workload aware elasticity for NoSQL. In: Eighth Eurosys Conference 2013, EuroSys 2013, Prague, Czech Republic, 14–17 April 2013, pp. 183–196 (2013)
28. Dutreilh, X., Rivierre, N., Moreau, A., Malenfant, J., Truck, I.: From data center resource allocation to control theory and back. In: IEEE CLOUD, pp. 410–417 (2010)
29. Dutta, S., Gera, S., Verma, A., Viswanathan, B.: Smartscale: automatic application scaling in enterprise clouds. In: IEEE CLOUD. pp. 221–228 (2012)
30. Elmore, A.J., Das, S., Agrawal, D., El Abbadi, A.: Zephyr: live migration in shared nothing databases for elastic cloud platforms. In: Proceedings of the 2011 ACM SIGMOD International Conference on Management of Data, pp. 301–312 (2011)
31. Elmore, A.J., Das, S., Pucher, A., Agrawal, D., El Abbadi, A., Yan, X.: Characterizing tenant behavior for placement and crisis mitigation in multitenant DBMSS, pp. 517–528 (2013)
32. Fernandez, H., Pierre, G., Kielmann, T.: Autoscaling web applications in heterogeneous cloud infrastructures. In: 2014 IEEE International Conference on Cloud Engineering, pp. 195–204 (2014)
33. Galante, G., de Bona, L.C.E.: A survey on cloud computing elasticity. In: 2012 IEEE Fifth International Conference on Utility and Cloud Computing (UCC), pp. 263–270 (2012)
34. Gedik, B., Andrade, H., Wu, K.L., Yu, P.S., Doo, M.: SPADE: the system s declarative stream processing engine. In: Proceedings of the 2008 ACM SIGMOD International Conference on Management of Data, pp. 1123–1134 (2008)
35. Gueye, S.M.K., Palma, N.D., Rutten, É., Tchana, A., Berthier, N.: Coordinating self-sizing and self-repair managers for multi-tier systems. Future Gener. Comp. Syst. **35**, 14–26 (2014)
36. Han, R., Ghanem, M., Guo, L., Guo, Y., Osmond, M.: Enabling cost-aware and adaptive elasticity of multi-tier cloud applications. Future Gener. Comp. Syst. **32**, 82–98 (2014)
37. Han, R., Guo, L., Ghanem, M.M., Guo, Y.: Lightweight resource scaling for cloud applications. In: 2012 12th IEEE/ACM International Symposium on Cluster, Cloud and Grid Computing (CCGrid), pp. 644–651 (2012)
38. Iqbal, W., Dailey, M.N., Carrera, D., Janecek, P.: Adaptive resource provisioning for read intensive multi-tier applications in the cloud. Future Gener. Comput. Syst. **27**(6), 871–879 (2011)
39. Jamshidi, P., Ahmad, A., Pahl, C.: Autonomic resource provisioning for cloud-based software. In: Proceedings of the 9th International Symposium on Software Engineering for Adaptive and Self-managing Systems, SEAMS 2014, Hyderabad, India, 2–3 June 2014, pp. 95–104 (2014)

40. Kalyvianaki, E., Charalambous, T., Hand, S.: Self-adaptive and self-configured CPU resource provisioning for virtualized servers using Kalman filters. In: Proceedings of the 6th International Conference on Autonomic Computing, ICAC 2009, 15–19 June 2009, Barcelona, Spain, pp. 117–126 (2009)

41. Kassela, E., Boumpouka, C., Konstantinou, I., Koziris, N.: Automated workload-aware elasticity of NoSQL clusters in the cloud. In: 2014 IEEE International Conference on Big Data, Big Data 2014, Washington, DC, USA, 27–30 October 2014, pp. 195–200 (2014)

42. Katukoori, V.K.: Standardizing Availability Definition. University of New Orleans, New orleans (1995)

43. Kaur, P.D., Chana, I.: A resource elasticity framework for QOS-aware execution of cloud applications. Future Gener. Comp. Syst. **37**, 14–25 (2014)

44. Kephart, J.O., Chess, D.M.: The vision of autonomic computing. IEEE Comput. **36**(1), 41–50 (2003)

45. Konstantinou, I., Angelou, E., Tsoumakos, D., Boumpouka, C., Koziris, N., Sioutas, S.: Tiramola: elastic NoSQL provisioning through a cloud management platform. In: Proceedings of the 2012 ACM SIGMOD International Conference on Management of Data, pp. 725–728. ACM (2012)

46. Li, Z., Zhang, H., Obrien, L., Cai, R., Flint, S.: On evaluating commercial cloud services: a systematic review. J. Syst. Softw. **86**, 2371–2393 (2013)

47. Lim, H.C., Babu, S., Chase, J.S.: Automated control for elastic storage. In: Proceedings of the 7th International Conference on Autonomic Computing, pp. 1–10 (2010)

48. Lorido-Botran, T., Miguel-Alonso, J., Lozano, J.A.: A review of auto-scaling techniques for elastic applications in cloud environments. J. Grid Comput. **12**(4), 559–592 (2014)

49. Marshall, P., Keahey, K., Freeman, T.: Elastic site: using clouds to elastically extend site resources. In: CCGRID, pp. 43–52 (2010)

50. Mastroianni, C., Meo, M., Papuzzo, G.: Probabilistic consolidation of virtual machines in self-organizing cloud data centers. IEEE Trans. Cloud Comput. **1**(2), 215–228 (2013)

51. Moore, L., Bean, K., Ellahi, T.: A coordinated reactive and predictive approach to cloud elasticity. In: The Fourth International Conference on Cloud Computing, GRIDs, and Virtualization, CLOUD COMPUTING 2013, pp. 87–92 (2013)

52. Moore, L.R., Bean, K., Ellahi, T.: Transforming reactive auto-scaling into proactive auto-scaling. In: Proceedings of the 3rd International Workshop on Cloud Data and Platforms, pp. 7–12 (2013)

53. Naskos, A., Stachtiari, E., Gounaris, A., Katsaros, P., Tsoumakos, D., Konstantinou, I., Sioutas, S.: Dependable horizontal scaling based on probabilistic model checking. In: CCGrid (2015)

54. Nguyen, H., Shen, Z., Gu, X., Subbiah, S., Wilkes, J.: AGILE: elastic distributed resource scaling for infrastructure-as-a-service. In: 10th International Conference on Autonomic Computing, ICAC 2013, San Jose, CA, USA, 26–28 June 2013, pp. 69–82 (2013)

55. Paraiso, F., Merle, P., Seinturier, L.: Managing elasticity across multiple cloud providers. In: Proceedings of the 2013 International Workshop on Multi-cloud Applications and Federated Clouds, pp. 53–60 (2013)

56. Paraiso, F., Merle, P., Seinturier, L.: soCloud: a service-oriented component-based PaaS for managing portability, provisioning, elasticity, and high availability across multiple clouds. CoRR abs/1407.1963 (2014)

57. Perez-Palacin, D., Mirandola, R., Calinescu, R.: Synthesis of adaptation plans for cloud infrastructure with hybrid cost models. In: 2014 40th EUROMICRO Conference on Software Engineering and Advanced Applications, Verona, Italy, 27–29 August 2014, pp. 443–450 (2014)

58. di Sanzo, P., Rughetti, D., Ciciani, B., Quaglia, F.: Auto-tuning of cloud-based in-memory transactional data grids via machine learning. In: Second Symposium on Network Cloud Computing and Applications, NCCA 2012, London, UK, 3–4 December 2012, pp. 9–16 (2012)

59. Serafini, M., Mansour, E., Aboulnaga, A., Salem, K., Rafiq, T., Minhas, U.F.: Accordion: elastic scalability for database systems supporting distributed transactions. PVLDB **7**(12), 1035–1046 (2014)

60. Serrano, D., Bouchenak, S., Kouki, Y., Ledoux, T., Lejeune, J., Sopena, J., Arantes, L., Sens, P.: Towards QOS-oriented sla guarantees for online cloud services. In: 2013 13th IEEE/ACM International Symposium on Cluster, Cloud and Grid Computing (CCGrid), pp. 50–57 (2013)

61. Shen, Z., Subbiah, S., Gu, X., Wilkes, J.: Cloudscale: elastic resource scaling for multi-tenant cloud systems. In: Proceedings of the 2nd ACM Symposium on Cloud Computing, pp. 5:1–5:14 (2011)

62. da Silva Dias, A., Nakamura, L.H.V., Estrella, J.C., Santana, R.H.C., Santana, M.J.: Providing IaaS resources automatically through prediction and monitoring approaches. In: IEEE Symposium on Computers and Communications, ISCC 2014, Funchal, Madeira, Portugal, 23–26 June 2014, pp. 1–7 (2014)

63. Tan, Y., Nguyen, H., Shen, Z., Gu, X., Venkatramani, C., Rajan, D.: Prepare: predictive performance anomaly prevention for virtualized cloud systems. In: 2012 IEEE 32nd International Conference on Distributed Computing Systems (ICDCS), pp. 285–294 (2012)

64. Tsoumakos, D., Konstantinou, I., Boumpouka, C., Sioutas, S., Koziris, N.: Automated, elastic resource provisioning for NoSQL clusters using tiramola. In: 2013 13th IEEE/ACM International Symposium on Cluster, Cloud and Grid Computing (CCGrid), pp. 34–41 (2013)

65. Uhlig, R., Neiger, G., Rodgers, D., Santoni, A.L., Martins, F., Anderson, A.V., Bennett, S.M., Kägi, A., Leung, F.H., Smith, L.: Intel virtualization technology. Computer **38**(5), 48–56 (2005)

66. Vaquero, L.M., Morán, D., Galán, F., Alcaraz-Calero, J.M.: Towards runtime reconfiguration of application control policies in the cloud. J. Netw. Syst. Manage. **20**(4), 489–512 (2012)

67. Varga, A., Hornig, R.: An overview of the OMNeT++ simulation environment. In: Proceedings of the 1st International Conference on Simulation Tools and Techniques for Communications, Networks and Systems & Workshops, p. 60. ICST (2008)

68. Vasic, N., Novakovic, D.M., Miucin, S., Kostic, D., Bianchini, R.: Dejavu: accelerating resource allocation in virtualized environments. In: ASPLOS, pp. 423–436 (2012)

A Survey on Software Tools and Architectures for Deploying Multimedia-Aware Cloud Applications

Christos Tselios[✉] and George Tsolis

Citrix - Bytemobile, Patras, Greece
{christos.tselios,george.tsolis}@citrix.com

Abstract. Multimedia-aware cloud is a novel cloud paradigm which addresses the overall framework needed for cloud infrastructure to effectively process multimedia services in a distributed fashion, provides Quality of Experience (QoE) provisioning for a broad spectrum of multimedia applications and facilitates all sorts of parallel processing schemes and adaptation methods for various types of end-user devices. The main purpose of this paper is to present some of the dominant platforms, software packages and application delivery tools and architectures that might help a multimedia-related application to be easily deployed, maintained and scale-up with as little limitations to performance and end-user QoE as possible.

Keywords: Cloud computing · Virtualization · Multimedia-aware cloud · Hypervisors · Containers · IaaS · PaaS · SaaS

1 Introduction

Cloud computing is probably one of the most successful computing paradigms ever to be introduced in computer science. Since user liberation from computer hardware always seemed like a groundbreaking idea, the notion of physical freedom in the virtual domain that this novel ecosystem of hardware and software resources introduced attracted a significant number of researchers from both industry and academia.

According to [1] cloud computing is a model for enabling ubiquitous, convenient, on-demand network access to a shared pool of configurable computing resources such as networks, servers, storage, applications, and services that can be rapidly provisioned and released with minimal management effort or service provider interaction. Such a sophisticated and innovative computational system has revolutionized the imminent way that users interact with their personal devices by reducing the overall client side complexity and the hardware requirements it demanded.

One certain type of cloud-based services, which has strongly benefited by the growth and vast improvement of cloud computing, is no other than multimedia-related applications. Paired with the introduction of Web 2.0 and facilitated by the technological leaps that occurred in the last years, multimedia services appeared to have almost dominated the Internet [2]. Multimedia processing such as image and video retrieval typically requires intensive computational resources especially when similar content needs to be

© Springer International Publishing Switzerland 2016
I. Karydis et al. (Eds.): ALGOCLOUD 2015, LNCS 9511, pp. 168–180, 2016.
DOI: 10.1007/978-3-319-29919-8_13

simultaneously delivered to a substantial number of Internet or mobile users. Having this intensive computation executed in the power-constrained mobile devices users own, might prove to be quite an error-prone approach. By using cloud-based multimedia services, consumers do not need to pay for costly computing devices but simply have all sorts of multimedia applications processed, rendered and stored on powerful cloud servers and pay for the utilized resources instead [3].

The overall growth of cloud computing along with the constant demand for better multimedia services and applications, lead to a plethora of architectural and software solutions aiming to dominate this exponentially growing market. Several companies offer series of products throughout the architectural spectrum of the cloud, from Hypervisor and Container-based virtualization technologies to overall Platform-as-a- Service (PaaS) and Infrastructure-as-a-Service (IaaS) products. This paper aims to present certain solutions in each of the aforementioned categories, specifically those available by key industry players and opensource consortiums. Based on the overall presentation, the reader will acquire an overview of the overall ecosystem, the available proprietary and opensource components for design and deployment of a multimedia-oriented cloud service along with some significant hints that might ease a potential architectural decision in the first place.

The rest of the paper is organized as follows. Section 2 contains the available Hypervisor and Container-based virtualization solutions that mostly facilitate an in-house private-hosted multimedia cloud. Section 3 describes certain PaaS architectures, which natively support multimedia applications, while Sect. 4 gives an insight on similar IaaS solutions that are considered multimedia-friendly. Finally Sect. 5 summarizes and concludes this survey.

2 Hypervisor and Container-Based Virtualization

A Hypervisor is a dedicated software or firmware component that is able to virtualize system resources by utilizing highly efficient and sophisticated algorithms, thus allowing multiple operating systems running on different Virtual Machines (VMs) to share a single hardware host [4]. The hypervisor is actually in charge of all available resources, which are allocated accordingly, making sure that all VMs operate independently without disrupting each other.

Container-based Virtualization is a server virtualization method in which the virtualization layer runs as an application within the operating system allowing the kernel to support several completely functional yet totally isolated user space instances called guests. In this approach guests share hardware resources in a more direct way without having the overhead of installing an operating system in each one. Performance is significantly improved since hardware calls are handled by a single operating system while guests are not subjected to any short of software emulation. In addition, container-based virtualization implementations capable of live migration can also be used for dynamic load balancing inside a cluster. The major drawback of this approach is the diminished flexibility since all guests need to have identical kernels to the host.

Three of the most significant hypervisor examples are Xen, VMware ESXi and Kernel-based Virtual Machine (KVM) while the most popular Container-based virtualization schemas and management software are Linux Containers, Docker, Rocket, Kubernetes and Mesos.

2.1 Xen

Xen [5, 6] is an open-source hypervisor consisted by a small software layer on top of the physical hardware that provides all necessary services for allowing multiple operating systems to be concurrently executed on the same underlying hardware. It introduces the notion of separate *domains,* which are VMs build on top of the hypervisor itself. The most privileged of those VMs with direct access to hardware, called *dom0* (domain zero), is created first and is used to initiate management tasks (i.e. create, discard, migrate, save, restore) and allowing access to I/O devices for all other VMs. One of the main advantages of Xen-based virtual machines is live migration between physical hosts without any availability loss or service interruption. During live migration Xen copies VM memory to the destination node and executes a certain synchronization process thus providing an illusion of seamless migration. Such an attribute might strongly benefit Multimedia clouds, which demand constant transformation and high availability under stress. Many recent enhancements of Xen focus in the area of GPU support, implemented via VGA pass-through [40], providing opportunities of leveraging graphics acceleration in multimedia (transcoding) applications.

2.2 VMware ESXi

VMware ESXi [37] is an enterprise-class hypervisor developed by WMware for deploying and serving virtual computers. It is categorized as type-1 hypervisor meaning that it is able to operate on bare metal infrastructure and includes all necessary components to do so, such as modified microkernel, known as *vmkernel* which directly handles CPU and memory utilization. Access to other hardware resources such as network and storage devices is enabled through specific modules most of which derived from modified versions of the same pieces of code used in the official Linux kernel. For facilitating the overall interface connection to all modules ESXi uses the *vmklinux* module, an intermediate emulation layer with direct access to the *vmkernel* itself. One of the main features of ESXi bare-metal hypervisor is its significantly small footprint, which only reveals a very small attack surface for malware and over-the-network threats, thus improving reliability and security.

2.3 Kernel-Based Virtual Machine

KVM is a free, open-source virtualization solution, which enables advanced hypervisor attributes on the Linux kernel. It consists of a loadable kernel module, *kvm.ko*, which facilitates the core virtualization infrastructure and a processor-specific module *kvm-intel.ko* or *kvm-am.ko* for Intel and AMD processors respectively [7]. Upon loading the aforementioned kernel modules, KVM converts Linux kernel into a bare metal hypervisor and leverages the advanced features of modern hardware, thus delivering unsurpassed performance levels [8].

2.4 Linux Container

Linux Container (LXC) is an operating-system-level virtualization environment, which allows a single Linux host to deploy and control multiple isolated Linux containers [9]. It is based on native kernel support for isolated namespaces along with *cgroups*, a kernel feature that handles resource usage for a collection of processes enabling resource limitation, prioritization, accountability and control. In particular, *cgroups* is designed to cooperate with the kernel in order to handle the CPU, memory, block I/O and networking demands of each process separately thus ensuring the aforementioned isolation of an overall application, including aspects such as process trees (through a separate process identifier allocation scheme), networking parameters, user IDs and even mounted file systems. In this way the container is able to execute native instructions to the whole spectrum of hardware resources without any special interpretation mechanism.

2.5 Docker

Docker [10, 11] is an open-source project that uses a custom container type to automate application deployment. Once heavily dependent not only to kernel virtualization capabilities but to *cgroups* and *namespaces* as well, Docker evolved recently to a more independent solution after introducing the *libcontainer* library. Using Docker, a hardware/platform-agnostic element as a universal method of container creation and management, the creation of highly distributed and lightweight systems which might operate on both localhost and cloud is going to be significantly more simple, allowing scaling up to be fast and precise [13], exactly the type of service a multimedia cloud success depends upon.

2.6 Rocket

Rocket (abbreviated as rkt) is a container runtime, or in order to be more precise, a command line interface (CLI) for running application containers in Linux [12]. Rocket is designed to be composable, fast and secure, currently offering integration with *init* systems such as *systemd* and *upstart*, compatibility with existing container software (such as Docker) and some advanced network configuration plugins along with KVM-based swappable execution engines. Rocket appears to have certain security enhancements over Docker since each container is launched independently, while in the later the parent process of all container process is the Docker daemon itself [13]. Therefore, for providing one additional level of security in all applications deployed in multimedia clouds, this container type seems to be marginally prevailing.

2.7 Kubernetes

Kubernetes [14] is an open-source system for managing containerized applications across multiple hosts in a cluster, which includes certain mechanisms for facilitating application deployment, scaling, scheduling and maintenance [15]. One of its main features is the introduction of Pods, defined as a collocated group of applications connected by a common context. This element, that also defines the smallest deployable unit that can be created, scheduled and managed, is a conjunction of several namespaces all of them having access to shared resources. Once Pods are created, the system continuously monitors their health as well as the state of the machine they are operating on. If a failure is detected, the system utilizes an API object caller Replication Controller, which automatically creates new Pods on a healthy machine. The replicated set of Pods might constitute an entire application, a micro-service or one layer in a multi-tier application [15]. Such level of granularity is ideal for multimedia clouds in order to obtain a high level of end user Quality of Experience by introducing all necessary services for network monitoring, transmission control and error correction that ensure seamless media delivery under complex network conditions.

2.8 Mesos

Mesos [16] is an open-source cluster management solution, which provides efficient resource isolation and sharing across distributed applications. According to [17] the main intention of the platform is to distribute a scalable and resilient core and define a minimal interface, which would allow cross-framework asset handling. A key characteristic of Mesos implementation is its ability to push control of task scheduling and execution to cooperating functions and entities. This strategy is considered crucial for allowing frameworks operating on top of it to implement diverse approaches to various cluster issues while in the same time Mesos retain its scalability and robustness by minimizing the rate of changes in code required for the system to remain up-to-date. Mesos uses a two-level scheduling mechanism where resource offers are made to frameworks (applications that run on top of Mesos). The Mesos master node decides how many resources to offer on each framework, while each framework determines the resources it accepts and what application to execute on those resources. This method of resource allocation allows near-optimal data locality when sharing a cluster of nodes amongst diverse frameworks [18].

3 Platform-as-a-Service Media Cloud Solutions

One of the major models for delivering cloud computing services is Platform-as-a-Service (PaaS), in which providers deliver a cloud-hosted virtual development environment along with the necessary solution stack, allowing customers to develop, run and manage applications without the complexity of building, configuring and maintaining the infrastructure typically associated with application development and launching. PaaS simply encapsulates a software layer and provides it as a service, making it usable as a solid foundation for higher-level service implementation [3]. This approach greatly benefits the overall design of Multimedia cloud since all related applications can be deployed easier, faster and with less layers of fine-tuning from the developer point of view. There are several PaaS solutions tailored for multimedia application deployment with the most suitable to be presented in the sections below.

3.1   Amazon CloudFront

Amazon CloudFront [19] is a Content Delivery Network (CDN) currently residing within the Amazon Web Services (AWS) framework [23], offering solutions via programmable API for video streaming, website acceleration, content download and traffic logging. Unlike other CDN platforms, CloudFront does not require customers to sign long-term or monthly usage contract but operates on a pay-as-you-go basis. In addition, it provides benefits to users with more predictable bandwidth usage given the fact that they agree on committing to certain delivery volumes per time unit. Although the core functionality of CloudFront is optimized to operate in parallel with other AWS services such as Amazon Elastic Compute Cloud (EC2) [20], Elastic Load Balancing [21] and S3 storage [22], customers are able to use it configured with services located outside the Amazon ecosystem, even inside private datacenters.

3.2   Google Application Engine

Google Application Engine [24] (often abbreviated as App Engine) is a PaaS offering that allows developers to build, execute and maintain applications on Google-managed infrastructure. Fully compliant to the PaaS model, App Engine restricts users from accessing the underlying datacenter, only providing a secure, sandboxed environment along with the necessary options for scaling-up and request distribution in order to meet all sorts of traffic demands. Compared to other PaaS solutions, App Engine provides more infrastructures for developers to write scalable applications, however it allows read-only access to the file system and can only execute pieces of code called from an HTTP request.

3.3   Microsoft Azure Media Services

Microsoft Azure Media Services (AMS) [25] is an extensive cloud-based platform that enables developers to build scalable media management and delivery applications. It is mainly based on REST APIs that facilitate upload, encode, store and package video or audio content in a secure way for live streaming and on demand delivery to a wide set of client equipment. When it comes to live and On-demand streaming, AMS allows scalable streaming for any size audience, just-in-time packaging using a variety of protocols (i.e. HTTP Live Streaming – HLS and MPEG-DASH), direct integration with Azure CDN for automatic provisioning and full cloud digital video recorder (DVR) workflow capabilities. However, AMS maintains its end-to-end proprietary nature, somehow discouraging adoption from low budget researchers and academic partners. Microsoft recently offered limited platform access to startups for free [26], but the whole spectrum of services remains relatively expensive for non-enterprise users.

3.4 OpenShift

OpenShift [27] is an open hybrid cloud service solution by Red Hat, based on the overall software platform currently open-sourced under the name OpenShift Origin [28]. One of the most interesting features of OpenShift solution ecosystem is automated provisioning and systems management inside the application platform stack. This enhances application scalability rendering them capable of meeting the growing user and feature demand much easier. Currently OpenShift seems to be an extremely stable and robust platform, however the need for constant improvement still persists. In the upcoming version of OpenShift Origin, OpenShift 3, an architecture re-design is imminent, providing support for Docker containers, the Kubernetes container management system [14] and a series of new extensions to accelerate application development and deployment. In addition, the concept of an application as a separate object is removed in favor of more flexible composition of "services", allowing two web containers to reuse a database or expose a database directly to the edge of the network. This new attribute will hugely benefit multimedia cloud since applications not only will be executed faster but also become less resource demanding in terms of computational speed.

3.5 CloudFoundry

CloudFoundry is an open source cloud computing PaaS solution mainly focused not only on delivering applications but also provide continuous integration maintenance and support. It is often described as the industry's open PaaS, provides a wide choice of clouds, frameworks and application services and by being open source attracts a broad community of researchers and contributors. When an application is deployed to Cloud-Foundry, an image is created for it and stored locally. This image is then automatically deployed to a special container called Warden for constant operation, which is deployed inside a VM. Since an identical procedure is executed for all applications, the system needs an internal controller entity to allocate resources for spinning up both VMs and containers. This internal controller called BOSH [29], was developed as an independent project to provision and deploy software over hundreds of VMs, is particularly well suited for large distributed systems and can also be used to deploy almost all sorts of software. This kind of scalability makes the particular solution attractive for multimedia cloud deployment since BOSH ensures VM functionality and also a dedicated load-balancing router is available for routing incoming requests to the correct application, in particular to one of the containers where the application is operating.

3.6 IBM Bluemix

IBM Bluemix [30] is an open-standard, PaaS solution for building, running and managing applications. It provides mobile and web developers access to enterprise-level software tools for integration and security. The platform is designed to host scalable, resilient apps and application artifacts that can both scale to meet all needs, and remain highly available and quick to recover from problems. This can be achieved by a basic component separation to those that track the state of interactions (*stateful*) and those that do not (*stateless*). In this way moving apps flexibly when needed to achieve scalability and resiliency becomes seamless and efficient thus avoiding service interaction. In general, Bluemix solution offers tailor-made cloud deployments by extending Cloud-Foundry technology, Docker container principals and OpenStack functionality, in an IBM-orchestrated environment where all instances are centrally managed. When it comes to facilitating multimedia cloud, IBM Bluemix uses a third-party solution, called Ustream [31] to provide online secure video streaming and broadcasting services.

4   Infrastructure-as-a-ServiceMedia Cloud Architectures

Infrastructure-as-a-Service (IaaS) is the cloud model that clearly demonstrates the difference between traditional IT approach and the cloud-based infrastructure service [1]. The consumer has the capability to provision processing, storage, networks and other fundamental computing resources, being able to deploy and run arbitrary software, including operating systems and applications. It allows consumers to tailor their requirements at a more granular level can also deliver basic or complex capabilities as a service over the Internet, therefore enabling pooling and sharing of hardware resources, such as servers or storage as well as peripheral devices like firewalls and routers [3]. All modern implementations of IaaS model that are presented in this section are generally consisted by a series of independent yet cooperating sub-modules, each with a specific role in the overall architecture. This modular approach not only facilitates constant module improvement in a parallel way but also renders new feature introduction easier and much more agile.

4.1   OpenStack

OpenStack [32] is a cloud operating system that controls large pools of compute, storage and networking resources throughout a datacenter, all managed over a dashboard that gives administrators total control while empowering all connected users to provision resources through a web interface. The overall system is consisted of mainly three separate modules, Compute, Networking and Storage, which are all accessible via the Open-Stack Dashboard and over dedicated APIs, thus allowing ad-hoc system reconfiguration and resource allocation. All OpenStack modules have certain legacy codenames marking the original project they derived from.

Compute. Compute module (codename Nova) is a cloud computing fabric controller implemented in Python, designed to manage and automate pools of computer resources. It is designed to scale horizontally and has the ability to work with all sorts of hardware setups, from bare metal to high performance computing configurations and the majority of the available virtualization and container technologies, such as KVM, Xen and LXC for minimal overhead and better performance.

Networking. Networking module (codename Neutron) provides pluggable, scalable, API-driven network and IP management and ensures that networking would never become a limiting factor or bottleneck in a constantly increasing environment where number of nodes, routing configurations and security rules may quickly escalate to over six-figure numbers. In such an environment traditional network management techniques fall short on providing a truly scalable and automated method of control, constantly supporting user's ever-growing expectation for flexibility with quicker provisioning. Neutron supports floating IPs that enable dynamic traffic rerouting, load balancing features, software-defined-networking (SDN) technology like OpenFlow [33] and a vast extension framework so that third-party network services can be seamlessly deployed and managed.

Storage. Storage module (codenames Swift - Cinder) is a scalable redundant storage system. Provided that many organizations now have a variety of storage needs with certain requirements for cost and performance, Swift supports both Object and Block storage. Object Storage (Swift) is not a traditional file system, but rather a distributed storage system for static data such as virtual machine images, photo storage, email storage, backups and archives. Having no central "brain" or master point of control provides greater scalability, redundancy and durability. It provides a fully distributed, API-accessible storage platform that can be integrated directly into applications or used for backup, archiving and data retention, rendering it an excellent solution for cost-effective, scale-out storage demands. Block Storage (Cinder) allows block devices to be exposed and connected to compute instances for expanded storage, better performance and integration with enterprise storage platforms. It is therefore better suited for performance sensitive scenarios such as database storage, expandable file systems, or providing a server with access to raw block level storage.

Dashboard. OpenStack Dashboard (codename Horizon) is the main graphical interface for accessing, provisioning and automating cloud-based resources. It provides an extensible design allowing third-party service integration and monitoring. Dashboard also allows branding for enterprise users and commercial vendors and is labeled as the most user-friendly method of OpenStack administration.

4.2 Apache CloudStack

Apache CloudStack [34] is considered one of the most technologically advanced opensource IaaS platforms, designed to manage and orchestrate pools of storage, network and computational resources. It works with a variety of hypervisors and hypervisor-like technologies such as KVM, Xen, VMware vSphere, LXC and bare-metal installation. By using CloudStack it is possible to setup an on-demand elastic cloud computing service and allow resource provisioning from administrators and end-users. The platform is able of managing a large amount of physical servers even when they are installed on geographically distributed datacenters, automatically configure network and storage settings for all deployments, all via a dedicated graphical user interface and the necessary REST-like API support. These characteristics encapsulate services such as VM template management and maintenance, routing, firewalling Virtual Private Networking (VPN), storage access and storage replication. In addition, CloudStack offers a number of solutions for availability increase, for instance a multi-node installation of the main management server over load-balanced infrastructure and horizontally scalable VM initiation capabilities. Last but not least, it provides an Amazon EC2 API translation layer to permit common EC2 tools utilization in a CloudStack-based deployment, a feature that greatly enhances cloud interoperability.

The basic deployment architecture of CloudStack involves two major components: the Management Server and the Cloud Infrastructure. In a simplistic approach, one might describe the Management Server as the orchestration entity of the overall setup, which it is possible to operate from a single node and the Cloud Infrastructure as the whole set of resources that are going to be divided for application usage over the particular cloud. However, since CloudStack-based IaaS solutions offer a tremendous amount of features, this simplistic approach somehow fails to describe the whole spectrum of possibilities that such a deployment is capable of, especially for multimedia cloud services and applications.

4.3   VMware vCenter Server

VMware vCenter Server [35] provides a centralized and extensive platform for managing virtual infrastructure implemented upon the core virtualization solution provided by VMware, vSphere [36]. Using vCenter Server, administrators are able to obtain simple and automated control over the virtual environment thus confidently deliver infrastructure. vSphere, the underlying virtualization layer which encapsulates ESXi hypervisor [37], operating between vCenter and bare metal, appears to be the basic element towards building an overall IaaS platform, however it is the advanced and highly proactive automated management features delivered by the vCenter that allow best-practice, end-to-end workflows to be implemented. Constantly delivering business-critical SLAs require heavy usage of the whole spectrum of the automated management features provided by vCenter, such as distributed resource scheduling, high availability practices and fault-tolerant operation. In addition, vCenter supports a variety of APIs, which allow integration of physical and virtual management tools for maximum flexibility. Certain key features of paramount importance, especially when deploying a multimedia cloud service, are vCenter's

- **Centralized Control and Visibility,** where all essential functions of the platform can be managed using single-sign on in a web client scheme.
- **Multi-hypervisor Management,** where simplified and integrated management of different hypervisor types is supported.
- **Dynamic Resource Allocation,** where utilization across resource pools is constantly monitored and all available resources are redistributed amongst VMs, according to predefined rules based on operational needs and changing priorities.

Despite all the excellent features that VMware offers in the vCenter Server suite, the proprietary nature along with the closed-garden mentality the company embraces renders this substantial software solution strictly enterprise-oriented.

4.4   Eucalyptus

Eucalyptus [38] is an open source software platform for building AWS-compatible private and hybrid clouds. Until recently the platform was marketed by the company

Eucalyptus Systems, until its later acquisition by Hewlett-Packard (HP) [39] and the product rebrand to HP Helion Eucalyptus. Eucalyptus' key objective is leveraging the existing hardware infrastructure in order to create a self-service private cloud within a company's premises. IaaS features and services are delivered by abstracting the available yet heterogeneous computational, networking and storage resources, thus creating an elastic resource pool that can dynamically scale up or down depending on the overall workload demands. In addition, after establishing a partnership with AWS, Eucalyptus maintains a high level of Amazon API compatibility, empowering users to shift workloads seamlessly over the two environments, public and private. The benefits span, from increased organizational agility and enhanced cloud security to less expensive infrastructure utilization, making Eucalyptus a popular solution amongst content providers who crave for AWS support and the robustness HP is able to offer. However, strongly depending on a sole cloud service provider, even a market leader such as Amazon, introduces a single point of failure hazard on delivering an application. Since multimedia clouds are quite demanding in terms of availability, even the slightest amount of downtime might lead to lower QoE score accumulation.

5 Conclusions

Multimedia-aware cloud is a novel cloud paradigm which addresses the overall framework needed for cloud to effectively process multimedia services in a distributed fashion, provides QoE provisioning for a broad spectrum of multimedia applications and facilitates all sorts of parallel processing schemes and adaptation methods for various types of end-user devices.

The existing cloud ecosystem appears capable of supporting the aforementioned service types, by offering a huge number of solutions covering all layers of modern cloud infrastructure architecture, from simple PaaS products with the underlying hypervisor and container technology, to overwhelmingly complicated yet end-to-end IaaS solutions. In addition, some of these solutions can be proprietary or open-source licensed, thus linking the overall cost of multimedia application delivery to architectural choices, most of which occur in the early design phases. The overall results of this survey can be summarized in Table 1.

The main purpose of this paper was to track down the most important requirements of multimedia-aware cloud and present some of the dominant platforms, software packages and application delivery tools and architectures that might help a multimedia-related application to be easily deployed, maintained and scale-up with as little limitations to performance and end-user QoE as possible. After reading this paper, architects, software engineers and researchers should become aware of most of the cloud-related solutions that could be considered, reviewed, and evaluated for deploying a multimedia-aware application. The results of this survey show that although there are several proprietary solutions that could be considered trustworthy and reliable, the cost factor is significantly high to be neglected. In addition, open-source software and all related platforms appear to have reached a certain level of maturity, that provided that the

Table 1. Summary of notable Hypervisor, Container, PaaS and IaaS solutions used for deploying Media-aware Cloud Services

Hypervisors	Type	Key Contributors	License Type	Supported by
Xen	Bare Metal	Community, Citrix, Google, Intel	GNU GPL v2	The Linux Foundation
ESXi	Bare Metal	VMware Inc.	Proprietary	VMware Inc.
KVM	Native	Open Virtualization Alliance	GNU GPL v2	Red Hat, Canonical
Containers				
LXC	Container Soft.	Community	GNU GPL v2	Parallels, Google, IBM
Docker	Container Soft.	Community	Apache License	Docker Inc.
Rocket	Container Soft.	Community	Apache License	CoreOS
Kubernetes	Cluster Management	Community	Expat/MIT	Google, Docker
Mesos	Cluster Management	Apache Foundation	Apache License	Mesosphere
PaaS	**Type**		**Provider**	**License**
CloudFront	CDN Solution		Amazon	Proprietary
App Engine	Dev Environment		Google	Proprietary
Azure Media	Media Delivery Platf.		Microsoft	Proprietary
Openshift	Hybrid Cloud Service		Red Hat	Proprietary
CloudFoundry	PaaS software		Pivotal	Open Source
Bluemix	Open standard		IBM	Proprietary
IaaS	**Key Supporters**			**License**
OpenStack	AT&T, Red Hat, HP, Intel, Canonical, Citrix, Ericsson, Yahoo, NEC			Apache License 2.0
CloudStack	Citrix, Apache Foundation			Apache License 2.0
vCenter Server	VMware Inc			Proprietary
Eucalyptus	HP			Proprietary

necessary manpower and time availability exists, they may well be a cost-effective and absolutely noteworthy solution.

Acknowledgements. This paper was supported by the DIOGENES Project (GSRT/GR-IL 3274). The authors would like to thank all reviewers and members of the consortium for their comments and remarks.

References

1. Mell, P., Grance, T.: The NIST Definition of Cloud Computing. National Institute of Standards and Technology, Special Publication 800–145, September 2011
2. Tselios, C., Politis, I., Tselios, V., Kotsopoulos, S., Dagiuklas, T.: Cloud computing: a great revenue opportunity for telecommunication industry. In: 51st FITCE Congress (FITCE), Poznan, Poland, September 2012
3. Tselios, C., Politis, I., Birkos, K., Dagiuklas, T., Kotsopoulos, S.: Cloud for Multimedia applications and services over heterogeneous networks ensuring QoE. In: Proceedings of the IEEE 18th Computer Aided Modeling and Design of Communication Links and Networks (CAMAD) Workshop, Berlin, September 2013, pp. 94–98 (2013)
4. IBM: Hypervisors, virtualization, and the cloud: learn about hypervisors, system virtualization and how it works in a cloud environment. http://www.ibm.com/developerworks/cloud/library/cl-hypervisorcompare/cl-hypervisorcompare-pdf.pdf
5. Lee, M., Krishnakumar, S.A., Krishnan, P., Singh, N., Yajnik, S.: Supporting soft real-time tasks in the Xen hypervisor. In: Proceedings of the 6th ACM SIGPLAN/SIGOPS International Conference on Virtual Execution Environments (VEE 2010), pp. 97–108. ACM, New York
6. Xen Hypervisor. http://www.xenproject.org/
7. Kernel Virtual Machine. http://www.linux-kvm.org/page/Main_Page
8. OVA: KVM Overview. https://openvirtualizationalliance.org/what-kvm/overview

9. IBM: LXC: Linux Container Tools. http://www.ibm.com/developerworks/linux/library/l-lxc-containers/l-lxc-containers-pdf.pdf

10. Docker. https://www.docker.com/

11. Docker: Understand Docker Architecture. https://docs.docker.com/docker/introduction/understanding-docker/

12. CoreOS: CoreOS is building a container runtime, rkt. Available: https://coreos.com/blog/rocket/

13. IBM: Initial experiment and assessment of CoreOS Rocket. https://www.ibm.com/developerworks/community/blogs/1ba56fe3efad432fa1ab58ba3910b073/entry/initial_experiment_and_assessment_of_coreos_rocket?lang=en

14. Kubernetes. http://kubernetes.io/

15. KubernetesGithub Repository: User Documentation. https://github.com/GoogleCloudPlatform/kubernetes/blob/master/docs/overview.md

16. Apache Foundation: Apache Mesos. http://mesos.apache.org/

17. Hindman, B., et al.: Mesos: A Platform for Fine-Grained Resource Sharing in the Data Center. NSDI (2011). https://www.cs.berkeley.edu/~alig/papers/mesos.pdf

18. Apache Foundation: Mesos Architecture. http://mesos.apache.org/documentation/latest/mesos-architecture/

19. Amazon Web Services: CloudFront. http://aws.amazon.com/cloudfront/

20. Amazon Web Services: Amazon EC2. http://aws.amazon.com/ec2/

21. Amazon Web Services: Amazon Elastic Load Balancing. http://aws.amazon.com/elasticloadbalancing/

22. Amazon Web Services: Amazon S3. http://aws.amazon.com/s3/

23. Amazon Web Services. http://aws.amazon.com/

24. Google: App Engine. https://cloud.google.com/appengine/

25. Microsoft Corp: Azure. https://azure.microsoft.com

26. Microsoft Corp: BizSpark. https://www.microsoft.com/bizspark/

27. Red Hat, OpenShift Enterprise. https://www.openshift.com/

28. Red Hat, OpenShift Origin. http://www.openshift.org/

29. BOSH. https://bosh.io

30. IBM: Bluemix. http://www.ibm.com/cloud-computing/bluemix/

31. Ustream. http://www.ustream.tv/

32. OpenStack. http://www.openstack.org/

33. ONF: OpenFlow. https://www.opennetworking.org/sdn-resources/openflow

34. Apache Foundation: Apache CloudStack. https://cloudstack.apache.org

35. VMware Inc.: vCenter Server. http://www.vmware.com/products/vcenter-server/

36. VMware Inc.: vSphere. https://www.vmware.com/products/vsphere

37. VMware Inc.: vSphere Hypervisor 6.0. https://my.vmware.com/web/vmware/evalcenter?p=free-esxi6

38. Hewlett-Packard: HP Helion Eucalyptus. https://www.eucalyptus.com/

39. Hewlett-Packard: Eucalyptus acquisition. http://www8.hp.com/us/en/hp-news/press-release.html?id=1790521#.VZFVzWSqpBc

40. Xen Project: Xen VGA Passthrough. http://wiki.xenproject.org/wiki/Xen_VGA_Passthrough

An Overview of Methods for Handling Evolving Graph Sequences

Andreas Kosmatopoulos[1]([✉]), Kalliopi Giannakopoulou[2],
Apostolos N. Papadopoulos[1], and Kostas Tsichlas[1]

[1] Department of Informatics, Aristotle University of Thessaloniki,
Thessaloniki, Greece
{akosmato,papadopo,tsichlas}@csd.auth.gr
[2] Department of Computer Engineering and Informatics,
University of Patras, Patras, Greece
gianakok@ceid.upatras.gr

Abstract. Graph data structures constitute a prominent way to model real-world networks. Most of the graphs originating from these networks are dynamic and constantly evolving. The state (snapshot) of a graph at various time instances forms an evolving graph sequence. By incorporating temporal information in the traditional graph queries, valuable characteristics regarding the nature of a graph can be extracted such as the evolution of the shortest path distance between two vertices through time. Most modern graph processing systems are not suitable for this task since they operate on single very large graphs. In this work we review centralized and distributed methods and solutions proposed towards handling evolving graph sequences.

Keywords: Snapshots · Evolving graph sequences · Temporal graphs

1 Introduction

Modern times, have witnessed a rapidly expanding volume of data generated by significantly different types of sources. A substantial portion of the available data, such as data originating from social networks, citation networks, sensor networks and others [13], can be modeled into graph data structures. The vertices of these graphs represent the entities of each network while the edges express relationships between the different entities. As an example, in a graph corresponding to a social network, the vertices denote the users of the network and the edges signify the friend-relationships between them.

A common characteristic of most real-world networks is that they do not remain static and are constantly evolving. For instance, the state of Facebook on one day is different to its state on the following day since there have been new user accounts created and friendships formed or deleted. Other networks, such as citation networks, only grow larger as they move forward in time since, due to the network's nature, vertices and edges are only added and never deleted.

© Springer International Publishing Switzerland 2016
I. Karydis et al. (Eds.): ALGOCLOUD 2015, LNCS 9511, pp. 181–192, 2016.
DOI: 10.1007/978-3-319-29919-8_14

It follows that, there exists a range of queries that aim to provide further insight on the nature of each network by incorporating temporal aspects in the traditional graph processing methods. Some examples of these queries would be to determine the evolution of a graph's diameter, the shortest path distance of two vertices through time and the vertex degree distribution of a graph at different time instances.

By periodically collecting the state of a graph at various time instances we form an *evolving graph sequence*. Current centralized and distributed graph processing systems such as Pregel [14], Neo4j [15], Trinity [19], Giraph [7] and others focus on processing single and very large graphs without supporting temporal extensions to the typical graph processing queries. As a result, these systems are not inherently suitable for performing analysis on evolving graph sequences.

Most of the research conducted towards handling evolving graph sequences aims to exploit the commonalities that exist between a graph in different time instances in order to improve space or time efficiency. As an example, even though social networks are dynamic and change over time, the majority of the users and the friend relationships between them remain the same across multiple time instances. For that reason, a system that effectively handles evolving graph sequences should perform better compared to a single graph processing system that operates on the individual graphs of the sequence.

The work performed in the area is in an inceptive stage and thus we present solutions for both centralized and parallel or distributed approaches. Among the centralized methods is the FVF framework by Ren et al. [17] that groups the sequence graphs into clusters and operates on them. Another method was proposed by Koloniari et al. [10] and it is based on maintaining a log of operations (defined as deltas) that occur in the graph between various time instances and employing it to reconstruct the graph at a particular time instance. Caro et al. [2,4] proposed space-efficient methods that utilize compact and self-indexed data structures to reduce the total space cost. Finally, methods have been proposed [1,8,18,21] that index the sequence in a manner that permits the efficient evaluation of certain queries. In the parallel and distributed setting there have been two main methods proposed: The DeltaGraph system [9] is based on the principle of deltas and aims to efficiently store and retrieve the graph at specific time instances. Finally, the G* system [11,12,20] is a parallel graph database that focuses on taking advantage of the commonalities present between a graph in different time instances to store the sequence in an efficient manner.

The rest of the work is organised as follows. In Sect. 2 we provide formal definitions regarding graphs, evolving graph sequences and a general problem definition. In Sect. 3 we present centralized methods and in Sect. 4 we focus on the parallel and distributed approaches. Finally, we conclude our work in Sect. 5.

2   Definitions

In this section we will provide some basic definitions about the general problem setting. First, we will formally define evolving graph sequences and then move on to discuss about the different query types that can be performed with regard to evolving graph sequences. Without loss of generality we will focus on undirected graphs since directed graphs mostly follow the same principles.

Definition 1 (Evolving Graph Sequence). *We define an evolving graph sequence $\mathcal{G}$ to be a collection of snapshots $\mathcal{G} = \langle G_1, G_2, G_3, \ldots \rangle$. A graph snapshot $G_i \in \mathcal{G}$ where $G_i = (V_i, E_i)$, corresponds to the graph G at time instance i and is characterized by a set of vertices V_i and a set of incoming and outgoing edges E_i.*

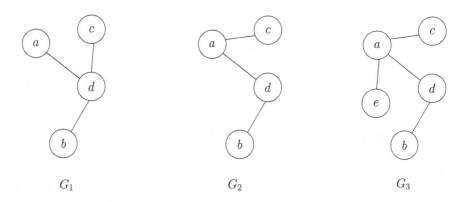

Fig. 1. An evolving graph sequence

The rate at which snapshots are obtained depends on the underlying network that the graph represents and is largely application-specific. Figure 1 depicts an evolving graph sequence which will serve as a running example for the remainder of this work. In this example, the evolving graph sequence $\mathcal{G}$ is composed of three snapshots G_1, G_2 and G_3 with each snapshot corresponding to the state of the graph G at time instances 1, 2 and 3 respectively. To obtain a particular snapshot from another snapshot in the sequence a set of operations has to be performed (e.g. by adding an edge between a and c in G_1 and removing the edge between c and d we obtain G_2). It is worth noting that a set of vertices or edges may not change at all in the entire sequence (e.g. b) and this fact can be exploited to reduce the total space or time cost when storing or querying the sequence respectively.

The aim of a system that handles evolving graph sequences is to efficiently store or index the sequence so as to answer historical analytic queries. We distinguish between two versions of the problem setting. In the *offline* version the entire sequence $\mathcal{G}$ is known beforehand and update operations are not supported in any snapshot (i.e. $\mathcal{G} = \langle G_1, G_2, G_3 \rangle$ only consists of G_1, G_2 and G_3 and no new snapshots are created). In the *online* version, $\mathcal{G}$ is constantly evolving and is not characterized by a "final" snapshot (i.e. $\mathcal{G} = \langle G_1, G_2, G_3, \ldots \rangle$ may eventually end up with more than three snapshots).

2.1   Query Types

The queries that can be performed upon evolving graph sequences can be characterized with respect to two main types [10]: their time domain and their graph scope. Regarding the time domain, queries are performed on either a particular time point or a time interval. In the case of a time point query we are interested in extracting a characteristic of the graph at a time instance t, while in a time interval query the objective is to study the evolution of a graph measure through an interval of time $[t, t']$. Queries are also distinguished by the scope of the graph that they operate on. More specifically, a query is focused on evaluating a graph measure concerning either a small set of vertices or the entire graph.

Most of the queries can be mapped to a combination of these two categories. As an example, consider the query "How has the shortest distance between a and c evolved over time instances t_s and t_e" which can be defined as a time interval query that focuses on a set of vertices. Similarly, the query "What is the diameter of G at time instance t_i" is a time point query that is concerned with the entire graph.

3   Centralized Methods

Having provided the basic definitions with respect to the problem of handling evolving graph sequences, we move on to methods and solutions proposed for centralized environments. We begin with the FVF framework by Ren et al. [17] followed by the works of Koloniari et al. [10] and Caro et al. [4]. We conclude the section by discussing indexing methods for evolving graph sequences that tackle certain historical queries.

3.1   The FVF Framework

The first centralized method we review is the FVF (FIND - VERIFY - FIX) framework proposed by Ren et al. [17]. The authors describe a method that consists of two phases, a preprocessing phase and a query-processing phase, and additionally propose storage models for the evolving graph sequences that support the aforementioned framework.

In the preprocessing phase the initial snapshots of the sequence are grouped into smaller clusters of similar snapshots. This is performed by defining a graph similarity measure and by incrementally adding snapshots in a cluster (starting from the first snapshot in the sequence) until a graph similarity threshold has been surpassed. At that point, a new empty cluster is created and the above procedure is repeated until all the snapshots have been examined. For each cluster, two representative graphs $G_\cap$ and $G_\cup$ are extracted which are the largest common subgraph and the smallest common supergraph of all snapshots in the cluster respectively. For example, if we assume that G_1 and G_2 from Fig. 1 are grouped in the same cluster their respective $G_\cap$ and $G_\cup$ graphs correspond to the graphs in Fig. 2.

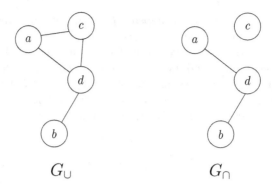

Fig. 2. $G_\cup$ and $G_\cap$ for the evolving graph sequence in Fig. 1

In the query-processing phase the authors use the clusters and their representative graphs to answer shortest path and closeness centrality queries. At first they evaluate the solution to a query for the representative graphs of the cluster ("FIND" step) on the basis that the solution will readily apply to a number of the snapshots in the cluster. In the "VERIFY" step, the evaluated solution is tested with each snapshot in the cluster in conjunction with a set of intuitive lemmas. For each snapshot that the evaluated solution does not apply, the framework attempts to "FIX" the solution so that it also applies to the aforementioned snapshot.

The authors also propose three storage models that can be used along with the FVF framework. The models make use of the similarities exhibited between successive snapshots and between representative graphs of successive clusters to reduce the total space cost of the evolving graph sequence. Finally, they assess their work through extensive experiments on both real and synthetic datasets.

3.2   Using Graph Deltas for Historical Queries

The authors in [10] advocate the use of graph deltas to support historical queries on evolving graph sequences. They begin by stating the operations that are supported on each snapshot, namely, $addNode(u_i)$, $remNode(u_i)$, $addEdge(u_i, u_j)$, $remEdge(u_i, u_j)$ which correspond to the addition or removal of a vertex u_i and the addition or removal of an edge between two vertices u_i and u_j respectively. Graph deltas are defined to be sets of such operations that when applied on a particular snapshot they yield another snapshot of the sequence. For example in Fig. 1, G_3 can be obtained by applying $\{addNode(e), addEdge(a, e)\}$ to G_2.

Furthermore, they define complete deltas to be sets of operations that when applied on the first snapshot of the sequence they are able to yield any of the sequence's snapshots.[1] Additionally, inverted deltas are defined to be sets of

---

[1] Certain snapshots require applying only a subset of the operations in a complete delta.

operations that when applied on a snapshot G_t they yield a snapshot $G_{t'}$ where $t' < t$, that is, $G_{t'}$ occurs "earlier" in the sequence than G_t.

Having defined the different types of deltas, the authors discuss snapshot materialization techniques and policies. More specifically, while any of the sequence's snapshots may be reconstructed if a complete and invertible delta and another one of the sequence's snapshots are maintained, it may be to the method's benefit to also maintain interposed snapshots to speed up snapshot materialization.

The next body of the work proposes three plans for efficient query processing. Perhaps the most universal of the proposed plans is a two-phase query plan that first materializes a particular snapshot according to the techniques discussed and then executes the query on the materialized snapshot. Finally, the authors discuss potential optimizations, delta indexing approaches and present some preliminary results of their solutions.

3.3 Compact Sequence Representations

Until this point the previous work we discussed was focused on reducing the total time cost of queries on evolving graph sequences. In the following works by Caro et al. [4] the authors address the problem of reducing the space cost when handling evolving graph sequences. Their proposed methods are heavily based on compact and self-indexed data structures that coupled with certain compression techniques (such as ETDC [3] and the PForDelta technique [22,23]) achieve overall high space efficiency with a good trade-off on the total time cost of the queries.

The authors use the concept of *contacts* as described by Nicosia et al. [16] to define temporal graphs.[2] A contact is defined to be a 4-tuple (u, v, t_s, t_e) that signifies the existence of an edge between vertices u and v during the time period $[t_s, t_e]$. The collection of all contacts is equivalent to the temporal graph itself, while, a particular snapshot G_t corresponds to the set of contacts (u, v, t_s, t_e) such that $t \in [t_s, t_e]$.

Next, operations that can be performed upon temporal graphs are presented. Those include:

- neighbor queries (i.e. report all neighbors of a vertex u),
- reverse neighbor queries (i.e. report all vertices that have a vertex u as neighbor),
- active edge queries (i.e. does there exist an edge between two vertices u and v at time instance t?),
- retrieving a snapshot of the graph at time instance t,
- edge state change queries (i.e. report all edges that have had their state changed at time instance t, that is all contacts that $t_s = t$ or $t_e = t$)

---

[2] Throughout the remainder of this work we will use the terms "evolving graph sequences" and "temporal graphs" interchangeably.

After a brief overview of the compression techniques and compact data structures they use in their work, the authors focus on the four temporal graph representations they propose along with their implementations that take advantage of the compression techniques. The first representation, called EdgeLog is an index that maintains for every vertex v in the temporal graph a list with the neighbors of v. Each neighbor of v is also equipped with a list containing all the time intervals that the particular edge exists in the sequence. The EdgeLog structure for a sequence composed by the graphs G_1 and G_2 of the example in Fig. 1, is depicted in Fig. 3.

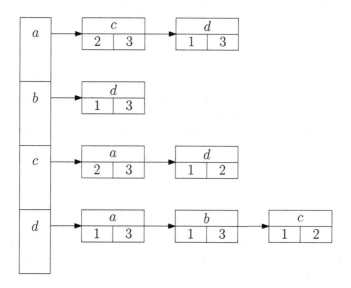

Fig. 3. EdgeLog and EveLog for the evolving graph sequence in Fig. 1

The second representation, called EveLog, follows a similar approach to the first. More specifically, EveLog is composed of a list with all the vertices that appear in the temporal graph. For each vertex v, there exists a list with all the "events" related to v (i.e. edge state change along with the vertex at the other end of the edge). The third representation is titled Compact Adjacency Sequence (CAS) and is based on the use of the Wavelet tree, while the fourth representation (CET) is based on the Interleaved Wavelet tree which is a data structure proposed in the same work as an additional asset to handling temporal graphs.

The work is concluded with extensive experimental evaluation over synthetic and real datasets through which the authors reach an interesting conclusion that there isn't a single best data structure for all the queries performed on temporal graphs. As a last remark, we should note that the above work focuses on the offline version of the problem, yet it also mentions alterations and modifications that need to be done in order for the solutions to apply to the online version.

3.4 Constructing Indices for Specific Queries

The work presented so far mostly focuses on efficiently storing, maintaining and retrieving the snapshots of an evolving graph sequences. There have been methods proposed in literature that instead aim to index the evolving graph sequence in a manner that permits the effective evaluation of specific queries. We present some notable examples in the section that follows.

Akiba et al. [1] describe dynamic indexing schemes that permit them to answer distance queries on either the last snapshot (current) or in any "older" snapshot in the sequence. Furthermore, they support the historical distance change-point query that reports all the time instances in the sequence where the distance between two vertices u and v changes. It is worth noting that in their work, they handle graphs that only support vertex additions and edge additions.

An other method that concentrates on answering shortest path queries was proposed by Huo et al. [8]. The authors make use of a Temporally Evolving Graph structure to store all the updates that occur in the sequence and proceed to use variations of Dijkstra's algorithm [5] to compute shortest paths. Furthermore, they speed up their solutions by making use of preprocessing indexes, namely, Contraction Hierarchies [6].

Yang et al. [21] propose an algorithm that discovers most frequently changing components in an evolving graph sequence. They begin by defining measures of change between vertices and the general problem of extracting the most frequently changing component and proceed to present their solutions.

Finally, Semertzidis et al. [18] tackle the problem of answering historical reachability queries. Their proposed index structure is called TimeReach and it is built in a manner that takes advantage of the strongly connected components that are present in a graph.

4 Parallel and Distributed Methods

In this section we turn our attention to methods and solutions that were proposed for parallel and distributed environments. The two systems that we will be analyzing are the DeltaGraph system by Khurana et al. [9] and the G* parallel graph database by Labouseur et al. [11,12,20].

4.1 The DeltaGraph System

Khurana et al. [9] designed and implemented a distributed system called Delta Graph that aims to efficiently store and retrieve snapshots from an evolving graph sequence. DeltaGraph supports time point (singlepoint) queries, time interval snapshot queries and multiple time point (multipoint) queries. Furthermore, along with the graph structure a query is also able to return the attributes of vertices and edges (e.g. name, weight etc.) The system is composed of two main components: the DeltaGraph index structure and the GraphPool in-memory data structure.

The DeltaGraph index is described as a rooted hierarchical graph structure that resembles a tree with adjacent leaves connected to each other in a bidirectional manner. The leaves of the structure correspond to snapshots of the sequence while the inner nodes correspond to graphs that can be obtained by applying a differential function (e.g. Intersection) to its children. The edges between the nodes store sets of deltas that are used to obtain a child node from its parent and they are horizontally partitioned between workers. It should be noted at this point that the only data stored are the sets of deltas and not the graphs themselves although the authors advocate the materialization of specific snapshots in DeltaGraph so as to speed up query time.

To answer a singlepoint query for a time instance t, the system locates through a binary search among the leaves the two adjacent leaves that "encompass" the query point t. Afterwards, it finds the minimum-weight path from the root to either of the two leaves, where the weight of an edge is set to be equal to the size of its respective delta. For multipoint queries, the system follows the same procedure with the difference being that instead of finding a path with minimum weight the system has to find the lowest-weight Steiner tree between the root and the multiple time instances.

The other component of the system is the GraphPool data structure which maintains in-memory a combination of materialized snapshots. More specifically, GraphPool maintains the current graph, historical snapshots and materialized graphs in a single combined graph. To determine which graphs contain a certain component or attribute the system makes uses of a mapping table. Finally, GraphPool is responsible for keeping the current graph index updated and cleaning up historical snapshots that are no longer needed.

4.2 The G* Graph Database

The last system we will be reviewing is the G* graph database by Labouseur et al. [11,20] that focuses on taking advantage of the commonalities that exist between snapshots in a sequence so that they are stored in an efficient manner.

In the G* system, each server is assigned a set of vertices along with all the outgoing edges of each vertex in the set. This achieves significant data locality since obtaining all of a vertex's edges can be accomplished without the need to contact any of the other servers. Furthermore, since the snapshots in a sequence exhibit similarities between them, G* avoids storing redundant information by only storing each version of a vertex once and, in that way, data that isn't modified between different snapshots isn't needlessly stored again.

Additionally, each server maintains an index named Compact Graph Index (CGI) that stores a single $(vertexID, disk\_location)$ pair for each vertex version that exists in a combination of the sequence's snapshots. For example, the CGI of a server maintaining vertex c of Fig. 1 would contain two pairs related to c: A pair for version c_1 in $\{G_1\}$ and another pair for version c_2 in $\{G_2, G_3\}$. It should be noted that the CGI has a low space overhead and can be mostly or fully kept in memory. As a last remark, the authors have proposed splitting the

Table 1. Summary of the works reviewed

Citation	Setting/Environment	Purpose/Approach
[17]	Centralized	Snapshot Storage & Retrieval, Shortest Paths, Closeness Centrality Queries
[10]	Centralized	Snapshot Storage & Retrieval, Two-Phase Query Plan
[4]	Centralized	Snapshot Storage & Retrieval, Compact and Self-Indexed Data Structures
[1]	Centralized	Historical Distance Queries
[8]	Centralized	Shortest Path Queries
[21]	Centralized	Discovery of Most Frequently Changing Components
[18]	Centralized	Historical Reachability Queries
[9]	Distributed	Snapshot Storage & Retrieval
[11]	Distributed	Snapshot Storage & Retrieval

CGI in a specific manner when a large number of graph combinations has been formed in its contents.

In a similar spirit to the storage module of G*, the CGI can also be used with regard to query processing to ensure that each version of vertex or edge is only processed once per query evaluation. Furthermore, the G* system supplies three types of primitives that can be used to construct graph query operators: summaries, combiners and bulk synchronous parallel (BSP) operators. Finally, in [12] the authors discuss snapshot replication and distribution techniques.

5   Conclusions

A significant fraction of contemporary networks can be modeled into graph data structures that are dynamic and constantly evolving. By integrating temporal information with typical graph queries we can obtain an improved understanding of a graph's overall nature. In this work we reviewed methods and systems proposed that aim to efficiently handle evolving graph sequences. A concise summary of the works presented can be seen on Table 1.

Acknowledgments. This research has been co-financed by the European Union (European Social Fund - ESF) and Greek national funds through the Operational Program "Education and Lifelong Learning of the National Strategic Reference Framework (NSRF) - Research Funding Program: Thales. Investing in knowledge society through the European Social Fund."

References

1. Akiba, T., Iwata, Y., Yoshida, Y.: Dynamic and historical shortest-path distance queries on large evolving networks by pruned landmark labeling. In: 23rd International World Wide Web Conference, WWW 2014, Seoul, Republic of Korea, 7–11 April 2014, pp. 237–248 (2014)
2. Brisaboa, N.R., Caro, D., Fariña, A., Rodríguez, M.A.: A compressed suffix-array strategy for temporal-graph indexing. In: Moura, E., Crochemore, M. (eds.) SPIRE 2014. LNCS, vol. 8799, pp. 77–88. Springer, Heidelberg (2014)
3. Brisaboa, N.R., Fariña, A., Navarro, G., Paramá, J.R.: Lightweight natural language text compression. Inf. Retr. **10**(1), 1–33 (2007)
4. Caro, D., Rodríguez, M.A., Brisaboa, N.R.: Data structures for temporal graphs based on compact sequence representations. Inf. Syst. **51**, 1–26 (2015)
5. Dijkstra, E.W.: A note on two problems in connexion with graphs. Numerische mathematik **1**(1), 269–271 (1959)
6. Geisberger, R., Sanders, P., Schultes, D., Delling, D.: Contraction hierarchies: faster and simpler hierarchical routing in road networks. In: McGeoch, C.C. (ed.) WEA 2008. LNCS, vol. 5038, pp. 319–333. Springer, Heidelberg (2008)
7. Apache Giraph: http://giraph.apache.org/
8. Huo, W., Tsotras, V.J.: Efficient temporal shortest path queries on evolving social graphs. In: Conference on Scientific and Statistical Database Management, SSDBM 2014, Aalborg, Denmark, June 30–July 02, 2014, pp. 38:1–38:4 (2014)
9. Khurana, U., Deshpande, A.: Efficient snapshot retrieval over historical graph data. In: 29th IEEE International Conference on Data Engineering, ICDE 2013, Brisbane, Australia, 8–12 April 2013, pp. 997–1008 (2013)
10. Koloniari, G., Souravlias, D., Pitoura, E.: On graph deltas for historical queries. In: WOSS (2012)
11. Labouseur, A.G., Birnbaum, J., Olsen, P.W., Spillane, S.R., Vijayan, J., Hwang, J., Han, W.: The G* graph database: efficiently managing large distributed dynamic graphs. Distrib. Parallel Databases **33**(4), 479–514 (2015)
12. Labouseur, A.G., Olsen, P.W., Hwang, J.: Scalable and robust management of dynamic graph data. In: Proceedings of the First International Workshop on Big Dynamic Distributed Data, Riva del Garda, Italy, 30 August 2013, pp. 43–48 (2013)
13. Leskovec, J., Krevl, A.: SNAP Datasets: Stanford large network dataset collection, June 2004. http://snap.stanford.edu/data
14. Malewicz, G., Austern, M.H., Bik, A.J.C., Dehnert, J.C., Horn, I., Leiser, N., Czajkowski, G.: Pregel: a system for large-scale graph processing. In: Proceedings of the ACM SIGMOD International Conference on Management of Data, SIGMOD 2010, Indianapolis, Indiana, USA, 6–10 June 2010, pp. 135–146 (2010)
15. Neo4j: http://neo4j.org/
16. Nicosia, V., Tang, J., Mascolo, C., Musolesi, M., Russo, G., Latora, V.: Graph metrics for temporal networks. In: Holme, P., Saramäki, J. (eds.) Temporal Networks, pp. 15–40. Springer, Heidelberg (2013)
17. Ren, C., Lo, E., Kao, B., Zhu, X., Cheng, R.: On querying historical evolving graph sequences. PVLDB **4**(11), 726–737 (2011)
18. Semertzidis, K., Pitoura, E., Lillis, K.: Timereach: historical reachability queries on evolving graphs. In: Proceedings of the 18th International Conference on Extending Database Technology, EDBT 2015, Brussels, Belgium, 23–27 March 2015, pp. 121–132 (2015)

19. Shao, B., Wang, H., Li, Y.: Trinity: a distributed graph engine on a memory cloud. In: Proceedings of the ACM SIGMOD International Conference on Management of Data, SIGMOD 2013, New York, NY, USA, 22–27 June 2013, pp. 505–516 (2013)

20. Spillane, S.R., Birnbaum, J., Bokser, D., Kemp, D., Labouseur, A.G., Olsen, P.W., Vijayan, J., Hwang, J., Yoon, J.: A demonstration of the g_* graph database system. In: 29th IEEE International Conference on Data Engineering, ICDE 2013, Brisbane, Australia, 8–12 April 2013, pp. 1356–1359 (2013)

21. Yang, Y., Yu, J.X., Gao, H., Pei, J., Li, J.: Mining most frequently changing component in evolving graphs. World Wide Web **17**(3), 351–376 (2014)

22. Zhang, J., Long, X., Suel, T.: Performance of compressed inverted list caching in search engines. In: Proceedings of the 17th International Conference on World Wide Web, WWW 2008, Beijing, China, 21–25 April 2008, pp. 387–396 (2008)

23. Zukowski, M., Héman, S., Nes, N., Boncz, P.A.: Super-scalar RAM-CPU cache compression. In: Proceedings of the 22nd International Conference on Data Engineering, ICDE 2006, Atlanta, GA, USA, 3–8 April 2006, p. 59 (2006)

Author Index